Praise for
eCarbon Card

Accelerating decarbonization demands inclusive action. This book by Sanjay Wahal compellingly argues for broader public participation and introduces the innovative eCarbon card system—a promising tool to empower individuals in tracking and reducing their carbon footprint. A timely and thought-provoking concept worth exploring to democratize climate responsibility and drive systemic change.

—**Mr. Sumant Sinha,** Founder,
Chairman & CEO of ReNew, India

I had the privilege of working with Sanjay during his time as a Technology Innovation Leader at Celanese Corporation, and I have always admired his ability to combine deep technical expertise with strategic vision. In *The eCarbon Card Blueprint*, Sanjay applies the same innovative thinking and systems approach to one of the greatest challenges of our time—climate change. His framework for digital carbon accountability is both practical and forward-looking, offering a model that can drive meaningful action at both individual and national levels. This book is a must-read for anyone seeking bold, actionable solutions for a sustainable future.

—**Peter E. Holmes, Ph.D.**
Former Senior Executive
Celanese Corporation

The eCarbon Card Blueprint by Sanjay Wahal presents a compelling and timely vision for integrating digital infrastructure with climate action. This study resonates deeply with India's Net Zero aspirations and comes at a time when such aspects are still evolving in the Indian context. Wahal combines technical rigor, policy insight, and systems-level thinking to the challenge of carbon accountability, offering a scalable framework grounded in India's unique digital capabilities. As someone working at the intersection of energy technologies, sustainability, and societal impact, I find this book to be an important contribution to the decarbonization discourse, especially for India and other emerging economies.

—**Ashish Garg, Ph.D.,**
Professor of Sustainable Energy Engineering,
Kotak School of Sustainability & The Kesavan Center,
Indian Institute of Technology, Kanpur, India

During my tenure leading the Product Development Team at Buckeye Technologies, I had the opportunity to collaborate closely with Dr. Sanjay Wahal, who made significant contributions in developing and commercializing airlaid absorbent cores for personal hygiene applications. His technical depth and strategic insight into product development set him apart. In *The eCarbon Card Blueprint*, Sanjay extends this systems-thinking approach to the urgent challenge of decarbonization and climate action—creating a technically sound and visionary framework that connects digital infrastructure with citizen behaviour and market mechanisms. This book bridges innovation and real-world impact, offering a model that could transform how individuals and systems engage with carbon responsibility.

—Jeffrey S. Hurley, Ph.D.
Former General Manager, Automotive Business Group,
Buckeye Technologies (later Georgia-Pacific)

I have had the privilege of knowing Sanjay Wahal not only through professional interactions but also as family—Sanjay's daughter is married to my son. I have long admired his integrity, technical rigor, and his ability to translate complex innovation into scalable, social-impact solutions. In *The eCarbon Card Blueprint*, Dr. Wahal brings together India's advanced digital ecosystem, behavioural economics, and market-based mechanisms to offer a citizen-centric blueprint for decarbonization. This is a timely and practical model that bridges individual action and national strategy—ideal for nations seeking inclusive and technology-enabled pathways to climate responsibility.

—Mr. Philippe Goetschel,
Former executive at Microsoft, Lotus, Symantec.
McKinsey & Company consultant and Honorary Consul of Switzerland

As someone deeply involved in advancing carbon utilization technologies, I find the eCarbon Card Blueprint to be a rare, timely and insightful contribution. Dr. Wahal's vision of combining digital identity systems with practical economic incentives and market-driven mechanisms for carbon accountability is both innovative and grounded in real-world applicability. It offers a scalable model that can align industrial operations, policy frameworks, and citizen engagement toward Net Zero. For anyone seeking practical, technology-driven carbon management, this book is an essential resource.

—Umesh Jayaswal, Ph.D.,
Industry Technology Expert in Carbon Utilization, Boston

The eCarbon Card Blueprint offers one of most compelling and viable frameworks for India's decarbonizaion journey. By leveraging the country's robust digital backbone-Aadhar, UPI, and scalable fintech platforms-Dr Wahal demonstrates how carbon accountability can be made practical, transparent, and citizen-centric. This concept is not just very innovative; it is actionable and highly relevant for India's ambition to lead the world in digital-driven climate solutions. Beyond India, this book can become a playbook for the countries worldwide striving to meet their net-zero goals while ensuring fairness and inclusivity."

—Alok Tandon, Ph.D., Chairperson
Joint Electricity Regulatory Commission
for Goa and Union Territories

Having known Sanjay since our IIT Kanpur days, I have always admired his commitment to applying technical expertise to societal challenges. In *The eCarbon Card Blueprint*, Sanjay has developed a transformative concept that bridges technology, policy, and human behaviour to tackle climate change. The idea of integrating India's digital infrastructure with market-based carbon mechanisms is both visionary and practical—offering a model that could be replicated globally. It is a refreshing, actionable approach to democratizing climate action.

—Manoj Mishra, Ph.D.

The eCarbon Card™ Blueprint

ISBN (paperback): 978-1-968919-02-3
ISBN (ebook): 978-1-968919-03-0

Armin Lear Press, Inc.
215 W Riverside Drive, #4362
Estes Park, CO 80517

The eCarbon Card™ Blueprint

Digital Solutions for Carbon Accountability

India as a Model

Sanjay Wahal, PhD

ARMINLEAR

I dedicate this book to my dear wife,
ANUPAMA,
and our lovely daughters,
SHWETA and ARPITA,
and to the next generation, whose future demands
not just innovation, but integrity, inclusion, and bold
collective imagination.

The climate crisis is not just a test of our science — it is a test of our compassion and our courage.

— Christiana Figueres, Costa Rican diplomat and author of *The Future We Choose: Surviving the Climate Crisis*

The best way to predict the future is to design it.

— Buckminster Fuller, American architect, systems theorist, writer, designer, inventor, philosopher, and futurist

Contents

Foreword

Ajay Mathur, Ph.D., 2nd Director General,
International Solar Alliance

Climate change is arguably the greatest challenge facing human-kind. All, and I mean all initiatives, including those in sectors as widely far apart as AI, biotechnology, and social media, are and will continue to be measured on the scale of their climate-change effectiveness. This is not an intuitive scale; we are continually inventing indices (sometimes *post facto*) – such as productivity per kg of carbon emissions – which will measure whether a proposed initiative is more carbon friendly than others, till we are, as a global society, able to make the transition to a carbon-free world.

Scales of comparative measurement, and tools for the rapid implementation of carbon-free measures, are emerging as the key drivers of the transition. I want to emphasize the need for rapid implementation. We have very little time to achieve global decarbonization. Global studies indicate that the world will need to reach net-zero emissions by 2050 to limit warming to 1.5°C, and by 2070 to limit warming to 2°C. In other words, the world has a 20-year timeframe – between 2050 and 2070 – to reach net zero emissions. This is a herculean task. Renowned climate scientist and policy analyst Vaclav Smil argues that the maximum rate of growth historically for a new energy source is about 9 percent per year; on the other hand, the International Energy Agency shows that renewable energy will need to grow at 15 percent per year for the world to reach net zero emissions by 2050.

The need to draw upon new means and new pathways – those that were not available when the last great global energy transition (when oil and gas replaced coal as the dominant source of energy) occurred – to accelerate the current and ongoing transition to renewable energy is obvious. The question is what will be these new means and new pathways?

Sanjay Wahal, in this book, provides an answer: that piggybacking on the digital revolution can also enable the zero-carbon revolution. And he proposes two unique concepts to enable this. The first is the allocation of a *per capita* share of the carbon budget to each and every citizen of the world so that he or she is the proud owner of a carbon allowance. And the second is to enable the use, verification, and trading of these allocated carbon allowances through digital means – the eCarbon Card™ which is the title of this book. And he draws on the digital experience of India, which shows how rapid implementation is possible. For example, the UPI platform (and the book describes the UPI platform in detail) enabled digital transactions to increase by about 49.9 percent per year (by volume) and 14.1 percent per year (by value) between 2021 and 2024. This suddenly shines a new light on the IEA's goal of a 15 percent per year required increase in renewable energy in order to meet the 2050 net-zero target.

The task does not seem herculean anymore!

There are those, of course, who will quibble over the fact

that an allocation of carbon allowances has been well near impossible to achieve in the global climate negotiations. And that linking verification of carbon-emitting activities with digitization has not occurred either. But I will argue that these two linked concepts provide us with a way out of the dilemma of time that faces us. And that by looking at the two concepts together, we could – if we politically want to – answer the big question of achieving carbon sustainability in our own lifetimes.

After all, as novelist Victor Hugo suggested in his writings, *"No power on earth can stop an idea whose time has come."* The ideas in this book are such ideas. They arrive not a moment too soon.

—Ajay Mathur, Ph.D.
2nd Director General, International Solar Alliance
Professor of Practice, School of Public Policy, Indian
Institute of Technology–Delhi

Former Director General: International Solar Alliance;
The Energy & Resources Institute; Bureau of Energy Efficiency
Member, Indian Delegation to the Climate Change
Negotiations, 2005, 2007–2012, and 2015

Preface

A Vision for Accountability in the Decarbonization Journey

As the world confronts the escalating consequences of climate change, the urgency for collective and innovative action has never been more apparent. Nations, organizations, and individuals are being called upon to redefine their relationship with the environment, rethink their consumption patterns, and contribute to a global effort to mitigate carbon emissions. Among these nations, India stands at a critical crossroads. As the world's third-largest emitter of greenhouse gases (GHGs), India faces the dual challenge of sustaining its rapid economic growth while addressing its environmental responsibilities. This task is monumental but not insurmountable — and it is precisely within this space of challenge and opportunity that the concept of the **eCarbon Card™** was born.

India's journey toward decarbonization is one of both necessity and possibility. India ranks 138th (among the top 10 countries) globally on Climate Vulnerability Index while its share of the global GHG emissions being only ~ 7%[1-4]. With its ambitious commitment to achieving net-zero emissions by 2070 and intermediate targets outlined under the Paris Agreement[5], India has positioned itself as a key player in the global fight against climate change. However, achieving these goals will require not just technological innovation and policy reforms, but also a fundamental shift in the way carbon accountability is embedded into the fabric of society. This transformation demands a mechanism that is not only equitable and scalable but also capable of empowering individuals and organizations to actively participate in the decarbonization journey.

The **eCarbon Card** concept was envisioned as a response to this need — a bold, actionable, and inclusive framework that democratizes carbon responsibility. This system moves beyond top-down policy mandates and engages citizens, businesses, and organizations in a participatory effort to manage and reduce carbon emissions. The eCarbon Card is designed to integrate seamlessly into daily life, making carbon accountability an inherent part of financial transactions, lifestyle choices, and business operations. It bridges the gap between high-level climate commitments and grassroots actions, enabling every individual and entity to become an agent of change.

At its core, the eCarbon Card is guided by a simple yet transformative idea: accountability. Climate change is a collective challenge, but its solutions must be individualized and localized. Every person's actions, every business's practices, and every government's policies contribute to the larger environmental picture. By tracking and managing carbon footprints at the micro level, the eCarbon Card empowers people to make informed, sustainable decisions and rewards them for their contributions to national and global climate goals.

The genesis of the eCarbon Card concept draws inspiration

from a blend of traditional and modern systems. India's Ration Card system, which has long been a cornerstone of equitable resource distribution, serves as a foundational model. This analog system, which has evolved into a sophisticated digital framework linked to Aadhaar[6,7], demonstrates how technology can be leveraged to create accountability and accessibility on a massive scale. Similarly, India's Unified Payments Interface (UPI)[8,9] showcases how digital infrastructure can transform everyday transactions, making them faster, transparent, and accessible to all segments of society. The eCarbon Card seeks to replicate and expand upon these successes, tailoring its approach to the unique demands of decarbonization.

As I reflect on the journey of conceptualizing the eCarbon Card, I am struck by the immense potential it holds — not just for India, but for the world. India, with its vast population, diverse socio-economic landscape, and robust digital infrastructure, is uniquely positioned to pioneer such a system. The Aadhaar ecosystem, which provides a digital identity to over 1.4 billion citizens, and the widespread adoption of UPI have created a foundation upon which the eCarbon Card can be built. These systems demonstrate that India has the capability to implement bold, large-scale initiatives that balance inclusivity, efficiency, and innovation.

However, the eCarbon Card is not merely a technical or policy-driven solution; it is a vision for fostering a cultural shift. It is about creating a society where sustainability is not just a goal but a way of life. By incentivizing low-carbon behaviors, promoting transparency in carbon accounting, and encouraging collaboration across sectors, the eCarbon Card aims to instill a sense of shared responsibility for the environment. It represents an opportunity to align economic growth with environmental stewardship, ensuring that development does not come at the expense of future generations.

The ideas presented in this book are the culmination of years of research, observation, and dialogue. They reflect my deep belief in the power of collective action and the critical role of individual accountability in addressing global challenges. This book is not just a guide; it is a call to action. It invites policymakers, businesses, and citizens to embrace the eCarbon Card concept as a pathway to a more sustainable future.

As you read through the chapters, you will see how the eCarbon Card system has been designed to address the unique challenges and opportunities of India's decarbonization journey. From its conceptual framework and technological requirements to its economic and environmental impacts, each chapter delves into the intricacies of making this vision a reality. The book also provides real-world examples, case studies, and actionable roadmaps to illustrate the system's viability and scalability.

My hope is that this book inspires readers to think critically about the role of accountability in climate action and to consider how innovative approaches like the eCarbon Card can drive meaningful change. Whether you are a policymaker, an entrepreneur, a researcher, or a concerned citizen, this book is for you. It is a blueprint for action, a source of inspiration, and a testament to the belief that together, we can build a sustainable future.

As I conclude this preface, I extend my heartfelt gratitude to the countless individuals, organizations, and thought leaders whose insights and contributions have shaped this work. The journey to decarbonization is a collective one, and it is through collaboration and shared vision that we will succeed.

Let us embark on this journey together — with accountability, empowerment, and hope lighting the way.

Note to International Readers

Though this book is structured around India as a case study, the eCarbon Card concept is inherently global in scope. India provides an ideal pilot setting: a vast and diverse population, a rising digital infrastructure (like Aadhaar, UPI, ONDC[10,11]), and a government deeply committed to sustainability goals. But the system described in these pages is not limited to any one country or culture.

Its core architecture — carbon budgeting, trading, and behavior-linked rewards — is adaptable to any economy in the world. Each country, city, or village can tailor this model to its energy mix, policy framework, and citizen engagement strategies. Whether you are a policymaker in Europe, a technologist in Africa, a sustainability officer in North America, or a community leader in Southeast Asia — this model can be localized, scaled, and co-owned.

Sanjay Wahal, PhD

Introduction

Path to Carbon Accountability for a Sustainable Future

The global climate crisis presents an unparalleled challenge, demanding urgent and innovative action, particularly from nations like India and China, whose rapid population growth and economic development contribute significantly to carbon emissions. While technological innovations offer viable pathways toward decarbonization, they cannot operate in isolation. Effective government policies must drive these efforts, promoting sustainable choices among citizens and businesses alike. Achieving decarbonization requires a synergistic top-down and bottom-up approach, where government initiatives are complemented by active citizen engagement.

The **eCarbon Card** concept provides a transformative vision for how nations can democratize decarbonization. By embedding carbon accountability into the financial transactions of individuals, businesses, and organizations, it creates a practical, scalable, and inclusive framework for tackling climate change. Unlike conventional top-level policies, the eCarbon Card empowers individuals and organizations to take responsibility for their carbon footprints. This innovative approach integrates environmental awareness into actionable steps, enabling everyone to contribute toward a nation's climate goals.

At its core, the eCarbon Card proposes a system where every citizen and corporate entity receives an annual carbon allowance or quota, fostering direct responsibility for carbon emissions. Those who exceed their allowances can purchase additional credits, while those who emit less can monetize their surplus through a dynamic carbon marketplace. The system emphasizes three foundational principles:

1. **Accountability**: Encouraging individuals and organizations to measure, manage, and mitigate their carbon emissions.
2. **Transparency**: Offering real-time tracking of emissions through a secure and accessible digital platform.
3. **Incentivization**: Rewarding low-emission behavior to make decarbonization both financially and socially beneficial.

The eCarbon Card system is more than just an ambitious concept; it is a blueprint for action that personalizes and democratizes the decarbonization journey. By linking carbon responsibility directly to individual and organizational behaviors, it fosters a culture of sustainability and environmental stewardship across all strata of society. The system enables individuals to make informed choices, businesses to optimize their operations, and governments to monitor and adjust climate strategies effectively.

Placing the eCarbon Card in the Global Framework

It is important to recognize that carbon accounting is not new. Globally, several respected frameworks already guide corporate, institutional, and national carbon reporting:

- The Greenhouse Gas Protocol (GHG Protocol)[12], co-developed by the World Resources Institute and the World Business Council for Sustainable Development, is the most widely used standard for emissions reporting.

- ISO 14064[13], developed by the International Organization for Standardization, sets global principles for quantifying and verifying GHG emissions.

- The UNFCCC National GHG Inventories[14-16] provide detailed emissions data at the national level, forming the basis for international climate negotiations and compliance under the Paris Agreement[1].

- Platforms like the Carbon Disclosure Project (CDP)[17,18] and Science-Based Targets initiative (SBTi)[19,20] help companies measure, disclose, and set targets in line with environmental, social, and governance (ESG) expectations.

These systems have played a crucial role in institutionalizing carbon responsibility. But they are largely limited to the organizational level — corporations, cities, and governments. They do not extend carbon literacy, decision-making power, or rewards to individuals or communities, particularly those in developing regions.

This is where the eCarbon Card fills the gap. It does not aim to replace existing protocols — it builds upon them. It pushes carbon accountability downstream, where behavior meets impact, where local decisions shape global outcomes.

Global Context: Countries Most Suitable for Implementation

The eCarbon Card system is particularly well-suited for countries with advanced digital infrastructure, strong governmental support for sustainability, or urgent environmental challenges.

- **Developed Nations with High Digital Literacy**: Countries like Sweden, Germany, Japan, and South Korea have robust digital ecosystems and high public awareness of climate issues. These nations are ideal for early adoption, as citizens are more likely to embrace innovative solutions like the eCarbon Card.

- **Countries with Existing Digital Identity Systems**: Nations such as India (Aadhaar), Estonia (e-Residency)[21, 22], and Singapore (Smart Nation initiative)[23-25] already have well-integrated digital identity systems, facilitating seamless integration of carbon tracking. For instance, India's Aadhaar infrastructure could enable rapid implementation and citizen participation.

- **Nations with Strong Governmental Support for Sustainability**: Northern European countries like Denmark, Finland, and the Netherlands, with their ambitious climate targets, offer conducive environments for pilot programs.

Countries Likely to Benefit Most from the eCarbon Card

India, as the world's third-largest emitter of greenhouse gases (GHG), stands to gain immensely from implementing an eCarbon Card system. With a robust digital infrastructure and growing urbanization, the country faces a dual challenge: sustaining economic growth while transitioning to a low-carbon future. For instance, pilot projects could begin in metropolitan hubs like Mumbai, Delhi, Bengaluru, and Chennai, where digital literacy and emissions are high.

Similarly, countries like Estonia, with its e-Residency program, and Singapore, with its Smart Nation initiative, possess the technological readiness and governance structures to adopt such systems effectively. Nordic nations like Sweden and Finland, with advanced environmental policies and digital penetration, are also strong candidates for large-scale rollouts.

India: The Ideal Launchpad for People-Powered Decarbonization

India stands at the crossroads of climate vulnerability and climate leadership. As one of the world's fastest-growing economies, its energy demand is expected to double by 2040, fueled by urbanization, industrial expansion, and a rapidly growing middle class. At the same time, India has committed to achieving net-zero emissions by 2070, a promise that calls for unprecedented innovation, scale, and citizen participation.

While this may seem like a daunting challenge, India possesses a unique set of assets that make it the ideal proving ground for the eCarbon Card system — a bold, inclusive, and scalable model for democratizing carbon accountability.

A Digital Infrastructure Like No Other

India has quietly built one of the most advanced digital public infrastructures in the world:

- Aadhaar provides a biometric digital ID to over 1.4 billion people.

- UPI processes over 10 billion transactions monthly, embedding real-time traceability into everyday spending.

- DigiLocker[26] and ONDC[10,11] enable digital access to personal records and commerce, respectively.

- The One Nation, One Ration Card (ONORC)[27] scheme has already demonstrated how digital tools can support portable, equitable access to essential resources across states.

These platforms can be seamlessly extended to track carbon allowances, behavioral incentives, and marketplace exchanges through the eCarbon Card. India's experience with ration cards evolving into smart, Aadhaar-linked tools shows that technology, when thoughtfully applied, can drive equitable and scalable delivery of public goods.

A Nation of Low-Carbon Citizens — Largely Unrecognized

More than 700 million Indians already live low-carbon lifestyles by necessity, not by design. They consume less energy, rely on shared transportation or walking, use seasonal and local foods, and contribute little to global emissions. Yet, they receive no recognition, benefit, or incentive for this frugality.

The eCarbon Card flips this narrative. It values and rewards low-carbon living, turning invisible sustainability into visible social and economic capital. It allows surplus credits from frugal users to be sold or traded to higher emitters, unlocking a new form of green income for rural and underrepresented communities.

Scalable for India, Adaptable for the World

India's federal structure, linguistic diversity, income disparities, and urban-rural divide mirror many global development challenges. If the eCarbon Card system can succeed here—balancing policy, incentives, and accessibility — it will offer a replicable model for:

- Emerging economies in Africa, Southeast Asia, and Latin America

- Climate-stressed nations seeking bottom-up decarbonization strategies

- Developed nations looking for scalable behavioral tools without political backlash

India's journey is not just national — it's a template for global transformation. In piloting the eCarbon Card, India has the opportunity to define a new climate compact: one that is inclusive, digital-first, and citizen-led.

The Broader Implications of the eCarbon Card

The eCarbon Card is not just a policy tool; it represents a societal shift toward sustainability. By integrating carbon accountability into everyday life, it bridges the gap between national climate goals and individual actions. This system also aligns with global trends, as seen in projects like Sweden's Climate Leap initiative and China's carbon trading markets.

Moreover, eCarbon Card can drive job creation, particularly in sectors like renewable energy, digital infrastructure, and data analytics. For instance, the system's implementation could generate up to a million new jobs in India, addressing unemployment while promoting green growth.

A New Climate Contract

This book does not propose a technological quick fix or a utopian future. It presents a practical roadmap — rooted in behavioral economics, policy innovation, and systems thinking — to make decarbonization democratic. It shows how we can move from carbon guilt to carbon agency, from regulatory compliance to everyday contribution.

By the end of this book, readers will understand how the eCarbon Card can help bridge climate ambition with community action, link carbon pricing with dignity, and turn emissions reduction into a national mission fueled by empowerment — not enforcement.

Let us imagine a future where every citizen is a carbon stakeholder, every action is a climate opportunity, and every nation has the tools to engage its people in the defining challenge of our time.

Structure and Purpose of This Book

The book begins with a **Preface**, sharing the author's motivation for writing this work and the urgency of rethinking how we approach personal carbon accountability. The **Introduction** offers essential context — laying the foundation for why India is uniquely positioned to lead a paradigm shift in decentralized, people-powered climate action.

From there, the book presents a comprehensive roadmap for conceptualizing, implementing, and scaling the eCarbon Card system — a transformative model for democratizing carbon responsibility and accelerating decarbonization.

- **Chapter 1** introduces the vision and distinctive value of the eCarbon Card concept.

- **Chapter 2** explores its structural foundations and operating principles.

- **Chapter 3** outlines the digital and technological infrastructure required to support the system.

- **Chapter 4** describes how carbon credit price is determined

- **Chapter 5** presents a phased roadmap for national and regional implementation.

- **Chapter 6** brings the concept to life through real-world applications and illustrative case studies.

- **Chapter 7** quantifies the potential economic, environmental, and social impacts of adoption.

- **Chapter 8** dives into the system of incentives, rewards, and governmental benefits to drive participation.

- **Chapter 9** candidly examines the challenges, risks, and limitations to be anticipated.

- **Chapter 10** provides a forward-looking view of long-term adaptability, innovation pathways, and policy evolution.

This is followed by a **Final Pitch** section — a high-level distillation of the concept designed to mobilize decision-makers, funders, and influencers. The **Epilogue** reflects on the broader implications of personal carbon accountability as a force for societal transformation.

Recognizing that any transformative concept — especially one as paradigm-shifting and multidimensional as the eCarbon Card — is bound to invite questions, critiques, and requests for clarification, this book includes a dedicated section titled **Critical Questions, Candid Answers**. Drawn from direct engagements with experts, policymakers, practitioners, academics, and community stakeholders who reviewed early drafts of the manuscript, this section addresses some of the most frequently raised and thought-provoking questions. These span a wide spectrum — from feasibility, equity, and behavioral incentives to market design, rural inclusion, digital access, and even philosophical debates surrounding personal carbon accountability. By curating and candidly responding to these concerns, the section serves not only as a reality check but also as a companion guide to the main chapters — deepening understanding, surfacing counterpoints, and reinforcing the conceptual integrity, adaptability, and policy relevance of the eCarbon Card. Above all, it is designed to promote clarity, transparency, and inclusive dialogue as we chart a path toward democratized decarbonization.

To support accessibility and shared understanding, this book concludes with a comprehensive **Glossary of Key Terms**, offering clear and concise definitions of core concepts, technical vocabulary, and acronyms used throughout the text. Given the interdisciplinary nature of the eCarbon Card — spanning climate policy, behavioral economics, digital identity systems, and carbon markets — this glossary is designed to serve both expert and non-expert audiences. It functions as a quick-reference guide for readers navigating the book's more technical sections, and ensures that foundational terms such as "carbon allowance," "digital twin," "CAT (Carbon Attribution Token)," and "consumption-based emissions" are consistently understood. The inclusion of this glossary underscores the book's commitment to clarity, transparency, and inclusive dialogue — empowering readers to confidently engage with the material and extend these ideas into real-world discussions and implementation.

Comprehensive **Appendices** include detailed explorations of blockchain integration, digital infrastructure requirements, carbon allowance methodologies for individuals and businesses, emissions tracking protocols, value chain integration, and realistic alternatives to blockchain. Additional sections provide a step-by-step implementation roadmap, UPI integration strategies, and national job creation estimates. Together, these Appendices serve as a practical reference toolkit — empowering readers to understand, explain, and advocate for the eCarbon Card in their own communities and institutions, while equipping policymakers, technologists, and entrepreneurs with the operational insights needed to take this concept from blueprint to reality.

By the end of this book, readers will gain a clear understanding of how the eCarbon Card can help reshape carbon economics, drive behavioral change, and unlock an equitable path to net-zero — not just in India, but around the world.

This is more than a policy proposal. It is a call to action for a livable, inclusive, and climate-secure future.

Chapter 1

The eCarbon Card System Concept:
Its Uniqueness, Impact, Benefits, and Vision

A New Compact for a Carbon-Constrained World

The eCarbon Card system is a pioneering initiative designed to transform how nations — starting with India — approach decarbonization in the 21st century. As one of the world's largest and most diverse emitters, India faces both an urgent climate imperative and an extraordinary opportunity to lead with innovation, inclusion, and scale.

At its core, the eCarbon Card introduces a structured, technology-driven framework for individuals and businesses to monitor, manage, and reduce their carbon footprints. By establishing personal and enterprise-level carbon wallets, the system assigns annual carbon allowances that reflect fair, science-based limits. These credits are dynamically adjusted and deducted based on lifestyle choices — transportation, energy use, food consumption, and more. Users who emit less than their allowance can save or trade their surplus, while those who exceed must purchase credits from others.

In doing so, the eCarbon Card transforms carbon accountability from an abstract, top-down concept into a tangible,
behavioral ecosystem that is bottom-up, participatory, and
real-time.

But this system is more than just a marketplace of emissions. It is a platform for carbon agency — one that recognizes
ordinary people as climate actors, not just passive consumers or
statistical units. It allows every citizen, household, entrepreneur,
and community to actively participate in the fight against climate
change, using accessible tools that reward sustainable living.

As the global race toward decarbonization intensifies, the
eCarbon Card offers more than a policy tool. It presents a new
social contract for the climate era — one that is inclusive, digital-
first, market-aligned, and rooted in the dignity and agency of
individuals. If successful in India, this framework could serve
as a global blueprint, showing how carbon responsibility can be
made equitable, empowering, and scalable across cultures and
economies.

The Need for a National Carbon Accounting System

India's emissions landscape reflects its growing energy demands,
urbanization, and industrial expansion, which together contribute significantly to the nation's carbon footprint. The unstruc-
tured and unaccounted emissions from key sectors such as
energy, transportation, agriculture, and manufacturing present
a formidable challenge to meeting global climate goals. These
emissions not only exacerbate climate change but also create
inefficiencies in energy use, leading to economic and environmental imbalances.

A national carbon accounting system such as the eCarbon Card is essential to address these inefficiencies. By tracking emissions at both individual and corporate levels, the system offers unprecedented transparency, enabling policymakers, businesses, and citizens to understand and address their environmental impacts. Unlike traditional emissions tracking systems, which focus primarily on industries, the eCarbon Card system is comprehensive, bringing every stakeholder — large and small — into the decarbonization effort. It not only identifies high-emission activities but also fosters a culture of actionable reduction, ensuring that sustainability becomes a shared responsibility across all sectors of society.

Moreover, with India committing to achieving net-zero emissions by 2070, this system aligns perfectly with national and international climate goals. It provides the necessary infrastructure to meet these targets systematically, while also ensuring that the socio-economic diversity of India is taken into account.

Understanding Carbon Credits and Allowances

The eCarbon Card system revolves around the twin pillars of **carbon credits** and **carbon allowances**, which together form the operational and behavioral core of the initiative. Carbon credits act as a tradable currency, representing measurable reductions in carbon emissions. These credits can be earned by individuals and businesses that adopt sustainable practices and stay within their allocated quotas. On the other hand, carbon allowances represent a cap on annual emissions. These allowances are carefully determined based on factors such as regional demographics, economic activity, and sector-specific emission benchmarks.

This dual framework incentivizes carbon-conscious behavior in several ways. Individuals and businesses who emit less than their allocated allowances can sell surplus credits on a carbon marketplace, creating a financial reward for sustainable practices. Conversely, those who exceed their quotas are required to purchase additional credits, introducing a financial disincentive for unsustainable behavior. By linking emissions reductions to economic benefits, the system motivates widespread participation and innovation, turning sustainability into a financially attractive proposition.

Additionally, the system ensures flexibility. Carbon allowances are progressively tightened over time to meet long-term reduction goals, while market dynamics in the carbon trading ecosystem ensure efficiency and adaptability.

Lessons from Global Precedents

The eCarbon Card system is inspired by global precedents that have demonstrated the effectiveness of structured carbon management systems. One notable example is the **European Union Emission Trading System (EU ETS)**[28], which operates on a cap-and-trade model. The EU ETS has successfully reduced emissions in energy-intensive industries by setting strict emission caps and enabling credit trading. The success of this model highlights the importance of a regulatory framework that combines stringent oversight with market-driven incentives.

Similarly, **California's Cap-and-Trade Program**[29] provides a compelling case for regional carbon markets. By fostering collaboration across industries and implementing cost-effective reduction strategies, California has become a leader in sub-national climate policy. Another valuable precedent is **China's National Carbon Market**[30], which, despite its complexity, demonstrates how phased implementation can integrate carbon trading into manufacturing-heavy economies.

While these models provide critical insights, the eCarbon Card system distinguishes itself through its inclusivity and personalization. By extending carbon accounting to individuals and incorporating advanced digital technologies, it bridges the gap between global climate policy and grassroots action. It offers a scalable model that ensures regional applicability and citizen engagement, making it uniquely suited to India's socio-economic realities.

This topic is delved into comprehensively in **Chapters 6 and 9**.

India's Unique Carbon Challenges and Opportunities

India faces distinct challenges in addressing its carbon emissions. The nation's energy sector is heavily reliant on coal, contributing significantly to greenhouse gas emissions. Rapid urbanization has also led to increased transportation emissions, while traditional agricultural practices in rural areas remain significant contributors to methane and other greenhouse gases. Furthermore, socio-economic disparities create unequal access to clean technologies, exacerbating emissions in underprivileged regions.

However, these challenges also present opportunities for innovation and systemic transformation. India's leadership in renewable energy — such as solar power under the International Solar Alliance — demonstrates its potential to lead global efforts in clean energy adoption. The eCarbon Card system capitalizes on these strengths by integrating India's existing clean energy initiatives into a comprehensive framework for emissions management. By addressing regional disparities, incentivizing green investments, and promoting energy-efficient technologies, the system aims to create a resilient and inclusive decarbonization pathway.

The Vision for the eCarbon Card

The eCarbon Card system envisions a future where every individual, business, and government entity is actively involved in reducing emissions. This vision is rooted in the principles of accountability, inclusivity, and economic viability. Unlike traditional top-down approaches to emissions reduction, the eCarbon Card empowers individuals to make informed, sustainable choices in their daily lives. Businesses, too, are incentivized to innovate and adopt cleaner technologies, creating a ripple effect of environmental and economic benefits.

At its core, the eCarbon Card system strives to:

- Enable comprehensive emissions tracking across individuals and corporations, ensuring accuracy and transparency.

- Promote accountability by tying emissions to tangible outcomes such as financial rewards and penalties.

- Provide actionable insights through data-driven recommendations, enabling participants to reduce their footprints effectively.

- Foster public awareness and engagement, creating a shared sense of responsibility for climate action.

- The eCarbon Card's vision aligns seamlessly with India's climate goals, offering a platform that not only supports decarbonization but also enhances economic resilience and social equity.

Key Features of the eCarbon Card System

The eCarbon Card system is built on a series of interlinked components that ensure its scalability and impact. Central to its design are personalized carbon quotas, dynamic digital accounts, and an integrated carbon credit marketplace. Personalized quotas allow for tailored emissions management based on individual and corporate needs, while digital accounts provide a seamless interface for tracking carbon footprints and managing transactions. The carbon credit marketplace incentivizes low-carbon behavior by turning unused allowances into financial assets.

Additionally, the system includes a robust reward and penalty structure. Individuals and businesses that adopt sustainable practices receive tangible benefits, such as tax rebates or discounts on eco-friendly products. Conversely, those exceeding their allowances face penalties, encouraging compliance and behavior change.

The Uniqueness of the eCarbon Card

What sets the eCarbon Card apart is its ability to democratize emissions management. By integrating carbon tracking into everyday life, it moves beyond industrial frameworks to create a system that is accessible to all. Its use of gamification[31], such as eco-leaderboards and rewards programs, enhances engagement and fosters long-term behavioral change.

Benefits of the eCarbon Card

The benefits of the eCarbon Card extend through environmental, economic, and social dimensions. It encourages sustainable practices, fosters economic opportunities through carbon trading, and supports India's climate goals. Moreover, its emphasis on transparency and equity builds public trust and promotes widespread participation.

Moving Beyond "Carrot and Stick"

In its initial framing, the eCarbon Card system emphasizes economic incentives — a reward-and-penalty loop grounded in behavioral economics. While this framework aligns well with reinforcement theory (trigger, behavior, reward), it must evolve to remain effective in the long term, especially across the diverse psychological and socio-economic landscapes of developing regions.

For the vast majority of the world's poor — many of whom are not major emitters to begin with — the idea of carbon footprints or global warming may feel distant, abstract, or irrelevant. In these contexts, extrinsic motivation alone (monetary rewards or penalties) may fail to create sustained behavioral change.

To address this, the eCarbon Card must integrate insights from Self-Determination Theory (SDT)[32], a widely validated psychological framework that identifies three essential human motivators:

1. **Autonomy** – the need to feel in control of one's actions
2. **Competence** – the need to feel capable and effective
3. **Relatedness** – the need to feel connected to others

These motivators, when activated, create intrinsic motivation — the kind that leads to long-term adoption, community pride, and scalable cultural shifts.

Designing for Intrinsic Motivation

Here's how the eCarbon Card can embed human-centered behavioral levers into its architecture:

1. **Relatedness: Community-Led Pride**
   - Local panchayats or ward-level dashboards showcasing carbon achievements
   - Recognition programs (e.g., "Low Carbon Champions")[33-35] for schools, neighborhoods, or cooperatives
   - Community carbon credits for collective actions like afforestation, clean cooking transitions, or waste segregation

Impact: Builds social capital, strengthens group identity, and normalizes sustainability norms.

2. **Autonomy: Ownership Over Impact**
   - Allow users to set personal carbon goals (e.g., reduce transport emissions by 20 percent in a year)
   - Provide multiple pathways for action (diet, transport, energy)

- Make the interface customizable and respectful of personal values and constraints

Impact: Reduces resistance, increases user satisfaction, and boosts self-efficacy.

3. Competence: Recognition and Feedback

- Instant visual feedback on performance (for example, "You've stayed 15 percent under your monthly allowance!")
- Access to tips, stories, and progress tracking over time
- Leaderboards, challenges, or milestone rewards for consistency and improvement

Impact: Reinforces confidence, sustains engagement, and encourages learning.

Reframing Rewards as Dignity, Not Dependency

In many rural and low-income contexts, the reward must go beyond currency. It must include recognition, agency, respect, and purpose. A farmer who walks instead of using a tractor, or a homemaker who shifts from biomass to LPG, may not care for abstract CO_2 metrics — but they do care about the wellbeing of their children, their standing in the community, and being seen as agents of positive change.

Thus, the eCarbon Card system must speak to hearts as much as wallets. It must not frame low-income populations as beneficiaries, but as co-authors of a new climate contract — where sustainability is not a sacrifice, but a source of identity and aspiration.

A Platform for Agency, Not Just Accounting

Where current climate systems tend to rely on guilt, compliance, or carbon taxes, the eCarbon Card creates a platform where individuals can:

- Visualize their impact in real-time

- Earn rewards — money and reputation — for sustainable living

- Participate in a national carbon economy as contributors, not bystanders

- Build local movements rooted in dignity and choice

This reframing is especially powerful in a country like India, where digital inclusion, social collectivism, and localized behavior patterns vary dramatically across contexts.

From Personal Footprints to Collective Transformation

The eCarbon Card system has the potential to become more than a carbon wallet. It can serve as a behavioral engine that powers India's decarbonization journey — and by extension, inspires similar frameworks globally. But for it to succeed, it must draw not only from economics and policy — but also from psychology, sociology, and systems thinking.

Incorporating intrinsic drivers alongside financial incentives will not only widen the card's reach — it will deepen its impact. It will allow us to build a climate movement that is bottom-up, culturally resonant, and emotionally intelligent.

Because at the end of the day, the climate is not saved by systems alone — it is saved by people who believe they have the power to shape it.

Challenges and Mitigation

While the eCarbon Card system holds immense promise, it must address challenges such as public acceptance, technological integration, and equity concerns. Transparent communication, robust infrastructure development, and targeted subsidies for vulnerable groups will be critical to overcoming these barriers and ensuring the system's success.

Conclusion

The eCarbon Card system represents a bold and innovative approach to managing emissions in a way that is inclusive, equitable, and scalable. By combining advanced technologies with robust policy frameworks, it empowers individuals and businesses to take meaningful action toward decarbonization. As a cornerstone of India's climate strategy, the eCarbon Card offers a transformative vision for a sustainable and resilient future.

Chapter 2

Mechanics and Structure of the eCarbon Card System

The eCarbon Card system is an ambitious, innovative framework designed to integrate carbon accountability into the daily lives of individuals, businesses, and organizations. At its heart, it transforms the abstract concept of carbon reduction into actionable, measurable, and incentivized behaviors. By leveraging technology, real-time monitoring, and a structured carbon credit marketplace, the system not only tracks emissions but also creates economic opportunities, rewarding sustainable choices.

The chapter explores the system's fundamental structure and operational mechanics, covering everything from its feasibility to how it assigns carbon allowances and tracks emissions. It delves into the system's feasibility, tailored features for diverse participants, and the technological infrastructure that ensures its smooth operation. Detailed examples illustrate how the system is applied to individuals and businesses while highlighting the role of a dynamic carbon marketplace. The eCarbon Card is designed not only to meet environmental goals but also to promote equity, accessibility, and economic resilience in

India's decarbonization journey. For readers interested in specific calculations, examples, and technical details, these have been included in various **Appendices**.

Before we delve into the details of the mechanics and structure of the eCarbon Card system, a review of a brief history of similar initiatives is in order.

A Brief History of Personal Carbon Card Initiatives

While the **eCarbon Card** may appear novel, the concept of individual carbon accounting has global roots. Several initiatives — governmental, financial, and academic — have attempted to translate emissions tracking into public-facing tools:

UK Personal Carbon Allowances[36-38] (2006)

Proposed by then–Environment Secretary David Miliband, the UK envisioned a universal carbon "credit card" for citizens. It would track emissions from travel, fuel, and electricity. Although never implemented, it introduced the idea of carbon as currency.

Climate Credit Card[39-41] (Switzerland, 2014)

Swiss company **Cornèrcard** introduced a credit card that calculated emissions per purchase and offset the carbon footprint of its users' spending automatically. It automatically tracked purchases and calculated the associated emissions, then investing in climate protection projects to offset those emissions. Cardholders have also received annual reports detailing their emissions and the offsetting efforts. Individuals could monitor and reflect on their personal carbon footprint based on real-life spending patterns. The project is recognized for exploring behavioral nudges, transparency, and digital feedback loops in voluntary carbon accountability — paralleling key features of the *eCarbon Card* concept.

DO Black Card by Doconomy[42,43] (Sweden, 2019)

In partnership with Mastercard, Swedish fintech Doconomy launched a credit card that integrated real-time carbon tracking with financial transactions, even including a carbon spending limit — and essentially capped user spending based on their carbon footprint. It served as a powerful early example of how personal carbon accountability can be integrated into digital finance — closely aligned with the behavioral principles behind the *eCarbon Card*. It was discontinued in 2022, partly due to adoption challenges.

Aspiration Zero Card[44] (USA, 2021)

The Aspiration Zero Card was a U.S.-based green credit card that rewarded users for climate-conscious spending by automatically offsetting carbon emissions through tree planting and offering cash-back incentives for sustainable behavior. Though it was ultimately discontinued in May 2023, it remains a key reference point for the growing trend of integrating environmental responsibility into consumer finance tools — a concept central to the *eCarbon Card* system.

West Bengal's Carbon Credit Card (India, 2023) [45,46]

Announced as the world's first state-backed personal carbon card, the West Bengal government proposed incentives for individuals to reduce carbon footprints. This pilot program, launched in July 2023, enables farmers to earn carbon credits for sustainable agricultural practices, such as reduced fertilizer use, zero tillage, and agroforestry. While in early-stage, it highlights India's openness to pioneering tools and marks a significant precedent in India for linking grassroots environmental stewardship with carbon finance, directly aligning with the principles envisioned in the *eCarbon Card* concept.

Academic Proposals: Theoretical Foundations of Personal Carbon Trading

Scholars across the world have explored personal carbon trading (PCT) systems as innovative mechanisms to engage individuals in climate action. In these models, citizens receive personal carbon budgets and are allowed to trade their unused emissions allowances with others — creating a behavioral and market-based approach to reducing carbon footprints[47-61].

This body of academic work—spanning economics, environmental policy, behavioral science, and systems design — has significantly informed the architecture of the eCarbon Card system, particularly in terms of fairness, scalability, and motivational framing.

These proposals emphasize:

- The feasibility of integrating carbon accounting into daily life

- The need for digital infrastructure and verification

- The importance of equity safeguards for low-income or marginalized users

- The potential economic benefits through carbon markets at the individual level

Appendix I – Overview of Academic Proposal: Theoretical Foundations of Personal Carbon Trading

lists pertinent references to the academic work along with key takeaways[53-61].

Feasibility of a Digital-Only Carbon Card System

India's rapid advancements in digital infrastructure form the foundation for implementing a fully digital eCarbon Card system. The country's Aadhaar-enabled identification framework, widespread smartphone adoption, and the success of the Unified Payments Interface (UPI) provide a robust, scalable, and accessible platform for carbon tracking and management.

The system envisions a digital interface accessible via smartphone apps and web portals where users, whether individuals or businesses, can monitor their carbon emissions in real-time and receive actionable feedback. Transactions linked to carbon emissions — such as fuel purchases, utility usage, or travel — are automatically recorded and deducted from the user's allowance. Businesses, on the other hand, integrate IoT-enabled tracking devices and supply chain reporting tools to manage emissions across their operations.

The integration of blockchain technology (see **Appendix II – What Is Blockchain and Why Does It Matter for the eCarbon Card**) ensures data security, transparency, and traceability, addressing concerns about tampering or manipulation. The digital-only nature ensures transparency, reduces administrative costs, and enhances scalability. Challenges such as public awareness, data privacy, and inclusivity are addressed through robust policies, secure platforms, and public outreach campaigns. Further technical details on the digital infrastructure have been provided in **Appendix III – Digital Infrastructure and Technological Framework for eCarbon Card System**.

Learning from What Didn't Work

Programs such as DO Black and Aspiration Zero showed that:

- Overly punitive systems risk disengagement

- Carbon data must be transparent, intuitive, and verifiable

- Users need to feel they are gaining control, not being monitored or taxed

The eCarbon Card incorporates these insights by embedding user agency, positive feedback, and choice-based flexibility.

Mechanism of the eCarbon Card System

A. Setting Carbon Allowances -
Carbon Allowances for Individuals

To meet India's climate commitments under the Paris Agreement[5] and subsequently agreed upon NDCs, individual carbon allowances are designed to limit annual emissions per person while ensuring equity. These can be calculated as follows -

Average Per Capita Allowance =
National Carbon Budget ÷ Population
Where the National Carbon Budget is set in alignment
with the NDCs.

As the population rises and the carbon budget drops, per-person allowances naturally decline.

With India's annual emissions in 2024 at approximately 2.75 billion tons of CO_2 and a per capita emission rate of 1.9 tons[62], the system proposes a progressive reduction in allowances beginning with the Year 2026 as follows -

1. **Baseline and Targets:**

Table 2.1

Projected Decline in Per Capita Carbon Allowance (India)

Year	Total Carbon Budget ($GtCO_2$)	Population (Billions)	Per Capita Allowance (tons CO_2)
2022	2.69	1.425	1.89
2024	2.75	1.451	1.90
2025	2.90	1.460	1.99
Phased Introduction of eCarbon Card System begins in 2026			
2026	2.60	1.500	1.73
2030	2.40	1.525	1.57
2035	2.05	1.579	1.30
2040	1.70	1.623	1.05
2050	1.00	1.680	0.60
2070	0.00	1.750	0.00

Fig. 2.1

eCarbon Card System Impact on Carbon Allowance and Total Carbon Budget

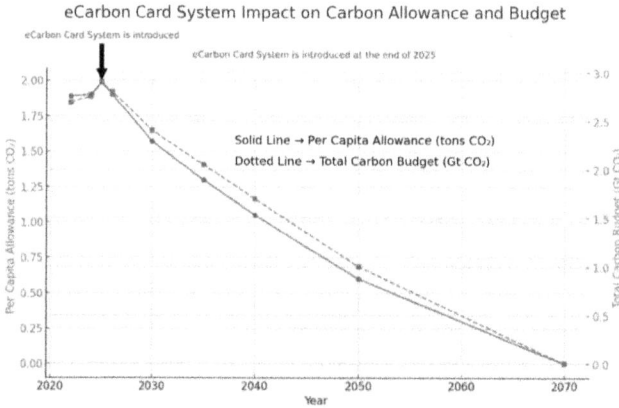

eCarbon Card System Impact on Carbon Allowance and Budget

Equity and Customization:

A foundational principle of the eCarbon Card system is fairness in carbon allowance allocation. But fairness is not self-evident — it must be defined, debated, and grounded in scientific, social, and ethical consensus. Fairness in carbon allocation cannot be left to market forces alone. It must be designed, debated, and constantly refined through a shared governance framework that is:

- Scientifically sound
- Socially just
- Technologically verifiable
- Economically adaptive

The carbon allowance framework for individuals and businesses is designed to reflect real-world disparities in lifestyle, infrastructure, and emissions capacity. However, to ensure legitimacy and wide acceptance, the development of these norms must involve a multi-stakeholder governance model that includes:

Government Participation

As the primary custodian of national climate targets and public policy, the government plays a central role in:

- Setting national carbon budgets (aligned with India's NDCs[63,64] – updated in 2023 and Net Zero commitments)

- Ensuring equity across states, income groups, and urban-rural divides

- Coordinating with ministries (MoEFCC, MNRE, MoHUA, etc.) for sectoral alignment

Academic and Scientific Input

Allocation norms must be backed by empirical evidence. Therefore, the system draws on:

- Behavioral and consumption data from NSSO, Census, and lifestyle surveys

- Life Cycle Assessment (LCA) models for emissions per product or service

- Research from academic institutions, think tanks (e.g.

CEEW, TERI, NITI Aayog), and international bodies
(e.g. IPCC guidelines)

A Carbon Allocation Advisory Panel composed of climate scientists, economists, and sociologists would be tasked with:

- Reviewing and updating allocation algorithms

- Validating emissions coefficients and regional disparities

- Setting adaptive guidelines to reflect changing behaviors or technologies

The system incorporates demographic and socioeconomic factors to ensure fair allocation of allowances:

- Urban residents may receive slightly lower allowances due to greater access to green infrastructure.

- Rural populations are allotted higher allowances to account for limited access to clean energy and public transport.

- Vulnerable groups, including large families and the elderly, receive even higher tailored adjustments to reflect their specific needs.

2. Adaptive and Participatory Governance

Recognizing that carbon budgets are not static, the eCarbon Card system is designed to be:

- Recalibrated annually based on actual usage trends and global targets

- Auditable by third parties (NGOs, community groups, independent reviewers)

- Flexible to incorporate new technologies, consumption shifts, or crises (e.g. energy shocks)

Additionally, a Public Feedback Mechanism allows citizens and businesses to:

- Submit concerns about fairness or data accuracy

- Participate in annual consultations on algorithmic updates

- Nominate innovations or practices that could merit allowance bonuses

3. Methodology for Individual Carbon Allowances

The methodology for calculating individual carbon allowances, detailed in **Appendix IV – Methodology for Allocating Individual Carbon Allowances**, uses data on lifestyle patterns, regional energy availability, and household size. By anchoring allowance norms in participatory, transparent, and data-driven processes, the eCarbon Card builds public trust, cross-sectoral legitimacy, and long-term resilience. A snapshot of the methodology is provided in Table 2.2 below.

Table 2.2

Factors Considered

Variable	Why It matters
Geographical Location	Urban vs. rural access to infrastructure, transport and energy
Household Size	Adjusts for shared vs. individual resource usage
Age and Health Status	Elderly or medically vulnerable user may need flexibility
Income Brackets	Acknowledges unequal ability to shift consumption behavior
Occupational Profile	High-mobility or remote regions considered accordingly

Carbon Allowances for Businesses

For corporations and organizations, carbon allowances are sector-specific and take into account their historical emissions, operational scale, and efficiency levels. These allowances are categorized into Scope 1 (direct emissions), Scope 2 (indirect emissions from purchased energy), and Scope 3 (supply chain emissions).

For enterprises, allowances are sector-specific and benchmarked based on:

- Historical emissions data (from annual sustainability or ESG reports)

- Industry norms (e.g. kg CO_2e per ton of steel or per product)

- Scope 1, 2, and 3 categorizations, as defined in the GHG Protocol[12]

- Operational size, revenue, number of employees, and asset intensity

Sector-Specific Considerations include -

- High-emission sectors, such as manufacturing, steel, and cement, receive higher initial allowances but are expected to reduce emissions at a faster rate.

- Service industries, with lower emissions, receive modest quotas but benefit from incentives for adopting renewable energy and efficient practices.

An example is provided in Table 2.3 below -

Table 2.3
Example Sector Profiles

Sector	Allowance Strategy
Cement / Steel	High initial budget, faster yearly reduction trajectory
IT / Services	Lower baseline, bonus for renewable usage & low Scope 3 emissions
Logistics	Bonus for route optimization, EV fleet adoption

Operational Integration

Corporations track their emissions using IoT-enabled monitoring systems that report energy consumption, waste generation, and transportation emissions. Advanced algorithms provide actionable insights, enabling companies to identify inefficiencies and optimize operations. For example, logistics companies implementing route optimization and electric vehicle fleets have reported significant emissions reductions and cost savings.

Companies that exceed their allowances must purchase additional credits, while those staying under their limits can monetize surplus credits in the carbon marketplace. Guidance on sectoral allowances and benchmarks is available in **Appendix V – Business Carbon Allowances and Sectoral Benchmarks**.

B. Tracking and Monitoring Emissions

Real-time tracking is the cornerstone of the eCarbon Card system, ensuring accuracy and accountability. The system integrates technologies like IoT, blockchain, and machine learning to collect and analyze emissions data across all activities.

Interfacing the eCarbon Card with UPI

UPI (Unified Payments Interface) could serve as the backbone for a carbon card system by linking carbon credits or allowances to financial transactions in real-time. Interfacing UPI (payment processing) with a carbon card system in India makes it possible to integrate carbon accounting with everyday financial transactions seamlessly by simplifying the tracking process. This integration would leverage UPI's existing digital infrastructure, which is already widely accepted and used across India, to track, manage, and incentivize carbon reductions through regular transactions.

The carbon card could be linked to each user's UPI account, allowing them to view their carbon balance alongside their financial balance within their banking apps. Whenever a user makes a purchase or pays for a service through UPI, the carbon footprint associated with that transaction can be automatically calculated and deducted from their carbon allowance. This integration would provide instant feedback on the environmental impact of each transaction and thus make it easy for users to keep track of their carbon allowances and adjust their behavior to stay within their personal carbon limits. For instance, a consumer purchasing groceries using UPI will see their carbon footprint calculated and deducted automatically, with low-carbon products incentivized through discounts or rewards.

Additionally, collaboration with banks, payment gateways, and digital wallet providers (like Paytm, Google Pay, and PhonePe) to integrate carbon tracking features might be considered.

For Individuals:

- Carbon usage is tracked via financial transactions, utility bills, and travel records. For example, purchasing gasoline deducts the corresponding emissions from an individual's carbon balance.

- Energy-efficient choices, such as using public transport or purchasing eco-friendly products, are rewarded with credits.

For Businesses:

- IoT devices installed in factories, warehouses, and transportation fleets monitor emissions in real time.

- Blockchain technology ensures that data remains tamper-proof, transparent, and traceable.

Further details on how a carbon card could track emissions across different sources are provided in **Appendix VI – Tracking Carbon Emissions**. Details on the technological infrastructure and its integration with the UPI payment system are provided in **Appendix VII – Integration with UPI and Digital Payments**.

C. Integration Across Value Chains

The eCarbon Card system extends its impact by integrating carbon tracking and accountability across the entire value chain, from raw material extraction to end-of-life disposal. For instance, manufacturing units adopting energy-efficient practices can earn credits, while consumers purchasing low-carbon products are incentivized through discounts. This holistic approach ensures that emissions are addressed comprehensively at every stage of production and consumption.

A detailed breakdown of approach to integration across value chains with examples is available in **Appendix VIII – Integration Across Value Chain**.

D. Incentivizing Sustainable Choices

The success of the eCarbon Card system depends on its ability to incentivize low-carbon behaviors effectively. These incentives span financial, social, and behavioral rewards. The eCarbon Card system uses a multi-faceted incentive structure to encourage carbon-conscious behaviors.

Financial Incentives:

- **For Individuals:** Cashback, discounts on eco-friendly products, and tax rebates motivate households to reduce emissions. For instance, families installing solar panels benefit from reduced utility bills and carbon credits that can be traded or saved.

- **For Businesses:** Tax deductions and credits reward companies that adopt green technologies, such as transitioning to renewable energy or reducing waste.

Social Recognition and Gamification[31]:

- Public awards, certifications, and gamified tiers (bronze, silver, gold) acknowledge and motivate sustained low-carbon behaviors.

- International examples, such as Japan's Eco Points program[65], highlight the effectiveness of such strategies in fostering a culture of sustainability.

E. Establishing a Carbon Credit Marketplace

The carbon credit marketplace forms the economic backbone of the eCarbon Card system, enabling the trading of surplus credits between low and high emitters. This dynamic ecosystem encourages innovation by rewarding low-emission technologies and practices and creates new revenue streams for individuals and businesses that adopt sustainable behaviors.

Dynamic Pricing:

Carbon credit prices are determined by supply and demand, with government intervention setting floors and ceilings to ensure market stability. During periods of high demand, credit prices rise, encouraging emissions reductions. Additionally, the government may set daily or monthly transaction caps or limits to prevent excessive trading and ensure that credits primarily benefit personal use, not commercial activity.

As carbon reduction goals tighten, the government may gradually increase the price of carbon credits to reflect higher national carbon reduction targets. For example, each year, the price floor could increase by a certain percentage to encourage individuals and businesses to lower their carbon footprint progressively. Credits tied to sectors with high emissions (like personal transportation) could carry higher costs, while credits for essential low-emission activities (like basic energy use) might be priced lower.

Chapter 4 goes in depth about the mechanism for the determination of carbon credit price.

Implementation of Trading Controls and Market Stabilizers:

Supply-Demand Regulation: Allow the regulatory authority to monitor market activity, setting price floors or ceilings to prevent speculative trading and ensure fair access

Transaction Limits: Set daily or monthly transaction caps to prevent excessive trading and ensure that credits primarily benefit personal use, not commercial activity.

Real-World Application:

Consider a household that conserves energy and accumulates surplus credits. They can sell these credits in the marketplace to a logistics company exceeding its quota, earning financial rewards while promoting sustainability. Examples of carbon credit pricing structure with government-set floor and ceiling prices and premium pricing with tiered structure are given in **Chapter 4**.

F. Dynamic Adjustments and Futureproofing

To remain effective and relevant, the system incorporates adaptive mechanisms:

- **Annual Reviews:** Allowances are adjusted based on technological advancements, policy changes, and feedback.

- **Emergency Provisions:** Short-term adjustments during extreme weather or economic crises balance human needs with environmental goals. Flexibility is built into the system to address these emergencies by allowing temporary adjustments in allowances to ensure that human needs are met without compromising long-term environmental goals.

The following Table summarizes the functional components of the System.

Table 2.4
Functional Components of the System

Component	Description
Carbon Wallet	Each user's digital wallet stores carbon credits and tracks debits from daily activities
Allowance Engine	Determines the yearly credit allocation based on user segment (region, income, occupation)
Transaction Tracker	Links carbon data to payment systems (UPI, card, kiosk) and logs emissions based on a national emissions factor database
Marketplace	Where users can sell excess credits, donate to community projects, or earn bonuses
Incentive System	Provides both monetary rewards and non-monetary recognition for low-carbon behavior
Community Dashboard	Aggregated, anonymized data showing local sustainability impact and collective progress

G. Leveraging GST (The Goods and Service Tax System)

The Goods and Services Tax (GST) system in India provides a unique GST Identification Number (GSTIN) to every business entity registered for taxation purposes. While GSTINs are generally not issued to individuals or households (since they are meant for businesses engaging in taxable transactions), the existing GST framework could serve as an inspiration or partial model for a carbon card system, even if it currently excludes individual, non-business-related emissions tracking. Here's how the GST system's infrastructure and methodology might facilitate a carbon card system and ways to potentially include individuals and households.

1. **Inspiration from GST's Centralized Tracking System**
   - The GST system has established a centralized digital infrastructure to track transactions and taxes across businesses, ensuring transparency and accountability. Similarly, a carbon card system could create a centralized platform for tracking carbon emissions across both businesses and individual activities.

   - GST's system is built to handle vast amounts of data, providing a backbone for real-time tracking and auditing, which is also essential for a carbon card system to be effective and scalable across the country.

2. Potential Link Between GST Infrastructure and Carbon Cards for Businesses

- For corporations and registered business entities, carbon tracking could leverage the existing GST network. Since companies already use GSTINs for tax compliance, the same infrastructure could potentially incorporate carbon credits or carbon emissions tracking.

- For example, a company's carbon footprint could be tracked based on the nature of goods and services they purchase, produce, and transport. Emission data could be linked to business transactions, which are already recorded for GST purposes.

3. Including Households and Individuals in the Carbon Card System

- Since GST primarily caters to businesses, a carbon card system would need a complementary approach for individual households. Below are a few potential approaches to include households and individuals in the system:

  » **Personal Carbon ID**: Like Aadhaar, a unique identification number could be issued to individuals and households specifically for carbon tracking. This could be a "Carbon ID" tied to each individual or household, regardless of their business activities.

» **Linking Carbon Tracking to Utility Bills**:
Carbon emissions tracking for individuals could
be facilitated through regular utilities (such as
electricity, water, and gas consumption) and fuel
purchases. Data from utility companies could be
linked to individuals' Carbon IDs, tracking their
usage patterns and estimating their carbon footprint
based on energy consumption.

» **Retail Partnerships**: For consumer goods
purchases, retail stores could link purchases
to individual Carbon IDs. For example, when
purchasing products with significant carbon
footprints (like fuel, meat, or high-energy
appliances), the transaction could be linked to the
individual's carbon card.

» **Incentivizing Data Sharing through
Voluntary Participation**: Unlike businesses,
individuals may not be obligated to participate.
However, offering incentives (for example,
rewards, discounts, or tax rebates) could encourage
individuals to participate voluntarily in the carbon
card system.

Conclusion

The eCarbon Card system represents a holistic, scalable approach to decarbonization blending advanced technology, equitable policy frameworks, and behavioral incentives to drive meaningful change. By combining real-time tracking, dynamic allowances, and financial incentives, it empowers individuals and businesses to take ownership of their emissions. Its integration with India's digital infrastructure ensures accessibility and scalability, while its marketplace encourages innovation and collaboration. The system's inclusive design, coupled with its potential to align economic growth with environmental goals, positions it as a critical tool in India's journey toward a sustainable future. As the eCarbon Card evolves, it offers a scalable model for other nations, underscoring India's leadership in innovative climate solutions.

Chapter 3

Required Technologies, Infrastructure, and
Roadmap for their Establishment

Introduction

Decarbonization is one of the most pressing challenges of our time, and the implementation of an innovative system like the eCarbon Card is a transformative step toward mitigating carbon emissions. However, achieving this vision requires a combination of advanced technologies, an extensive infrastructural network, and a meticulously planned roadmap. This chapter discusses the technological components, infrastructural needs, and phased strategies required to implement the eCarbon Card system, catering to the unique socio-economic diversity and climate goals of India.

System Design and Technological Framework

The eCarbon Card system's foundation lies in creating a centralized digital platform that is secure, scalable, and accessible to all stakeholders. This platform integrates various technologies to track emissions, manage carbon credits, facilitate transactions and provide actionable insights to users. It serves as the foundation for integrating diverse data sources, ensuring accuracy and accountability for individuals and corporations alike.

Digital Platform and Integration

The digital platform must function as the central hub for all activities, enabling seamless tracking of emissions across individual and corporate activities. This platform will integrate:

- **Carbon Tracking Systems**: Data collection from multiple sources, such as IoT-enabled smart meters, transportation apps, and utility bills, will allow the platform to calculate carbon footprints with high precision.

- **Credit Trading Mechanism**: A user-friendly marketplace will allow individuals and businesses to trade carbon credits, fostering economic incentives for sustainable behavior.

- **Data Analytics and Insights**: Leveraging machine learning, the system will provide real-time feedback and trends to help users make informed decisions.

This platform must be cloud-based to ensure it can scale to handle the data from millions of users and transactions. Additionally, integration with national databases like Aadhaar for individuals and PAN for corporations will ensure seamless identity verification.

Mobile Applications and Cloud Infrastructure

The eCarbon Card mobile application serves as the primary interface for users, providing real-time updates on carbon balances, transaction history, and actionable insights. The app must be:

- **User-Friendly**: Designed for ease of use across diverse demographics, with features like multilingual support and simple navigation.

- **Interactive**: Offering gamification elements, such as rewards for low-carbon activities, to engage users actively.

Cloud computing infrastructure is essential for storing and processing vast amounts of data. By leveraging cloud services, the system can be scaled to handle millions of users without compromising on performance or security.

Dynamic Pricing Algorithms

Dynamic pricing ensures the efficient functioning of the carbon credit marketplace. These algorithms adjust the price of carbon credit based on real-time supply and demand. For instance:

- **High Demand Scenarios**: If demand for credits exceeds supply, prices rise, incentivizing users to reduce emissions or adopt sustainable practices.

- **Surplus Credits**: In scenarios where users have excess credits, prices decrease, encouraging buyers to invest in green initiatives.

Dynamic pricing fosters a balanced and efficient market-place, rewarding sustainable behaviors while ensuring afford-ability for essential needs. This is further discussed in detail in the next chapter.

Key Technological Pillars Required

Key technological pillars of the system include blockchain for secure transactions, artificial intelligence for personalized rec-ommendations and IoT for real-time emission tracking. Each of these components contributes to the platform's reliability and adaptability, enabling it to meet the demands of a rapidly growing digital and environmental landscape.

A. Blockchain for Secure Transactions[66-69]

Blockchain technology (see **Appendix II**) forms the backbone of the eCarbon Card system's security and transparency. By utilizing blockchain, all transactions related to carbon cred-its — issuance, trading, and usage — are recorded immutably. This eliminates the risk of fraud, such as double-counting cred-its, while enhancing trust among users.

For example, each time a user purchases a carbon-inten-sive product, the system deducts the corresponding credits, recording the transaction on the blockchain. Peer-to-peer trad-ing of credits between users and corporations is also facilitated transparently, with smart contracts automating processes like price determination and compliance checks.

There are several examples worldwide where blockchain technology is being utilized for purposes similar to the concept of a carbon card, specifically for tracking, managing, and incentivizing sustainable practices and resource allocation. Here are some notable examples:

1. **IBM and Energy Blockchain Labs – Carbon Credit Trading in China**[70-72]

Purpose: IBM collaborated with Energy Blockchain Labs to create a blockchain-based platform specifically for carbon credit trading. This platform was designed to help businesses in China track and trade carbon credits in a transparent, efficient way.

How It Works: Blockchain is used to create a secure, immutable ledger where carbon credits can be issued, tracked, and traded. This ledger ensures that all transactions are transparent and verifiable, reducing the chance of double-counting emissions reductions or fraud.

Impact: The platform has helped streamline the carbon credit trading process, making it more accessible and trustworthy. This is somewhat similar to the carbon card system, as it uses blockchain to track emissions, enforce accountability, and create incentives for emission reductions.

2. **Power Ledger – Peer-to-Peer Energy Trading in Australia**[73-75]

Purpose: Power Ledger, an Australian-based blockchain platform, enables peer-to-peer (P2P) energy trading. This allows individuals and businesses to generate their own renewable energy (for example, solar) and trade any excess power directly with neighbors, businesses, or other entities in a decentralized manner.

How It Works: Power Ledger's blockchain records energy production, consumption, and trading transactions in real-time, enabling transparent and efficient trading. Individuals or entities with excess renewable energy can sell it to others, creating a decentralized energy marketplace.

Impact: This system provides financial incentives for individuals to invest in renewable energy, aligns with carbon reduction goals, and gives participants control over their energy usage. Although it's not a carbon card, the P2P trading approach and transparent tracking system align with the goals of reducing emissions and encouraging sustainable practices.

3. Swytch – Tokenized Carbon Credits for Sustainable Energy[76,77]

Purpose: Swytch uses blockchain to reward renewable energy producers by issuing tokenized carbon credits. Each time renewable energy is generated, Swytch issues digital tokens as rewards, which can be traded or redeemed.

How It Works: Using smart meters, Swytch verifies the amount of renewable energy produced. This data is then recorded on a blockchain, which issues tokens based on the carbon offset. The tokens incentivize renewable energy production, functioning similarly to how a carbon card might reward individual or corporate actions that reduce emissions.

Impact: Swytch has been effective in creating an incentive system for renewable energy production. The token system and blockchain provide transparency, traceability, and a tangible reward for sustainable actions, showing how blockchain can align economic incentives with environmental goals.

4. Everledger – Blockchain for Diamond and Supply Chain Tracking[78-81]

Purpose: Everledger uses blockchain technology to track the provenance of diamonds and other valuable goods throughout the supply chain, from source to end consumer. While not directly related to carbon tracking, it showcases how blockchain can be used to ensure ethical sourcing, which can apply to carbon tracking.

How It Works: Blockchain ensures that the origin and journey of diamonds are recorded immutably, preventing issues like the trading of conflict diamonds. This model can be applied to a carbon card system where emissions and sustainability data are recorded throughout a product's life cycle, giving consumers insight into the carbon footprint of their purchases.

Impact: Everledger has helped promote ethical sourcing and transparency in the diamond industry, proving that blockchain can effectively enhance trust and traceability in supply chains. This approach could similarly be adapted to track and verify carbon reductions in a carbon card system.

5. UK's Proposed Carbon Emission Trading System with Blockchain[82,83]

Purpose: In recent years, the UK has explored blockchain-based carbon trading systems to complement its cap-and-trade emissions scheme. The idea is to make carbon trading more efficient, transparent, and secure.

How It Works: By leveraging blockchain, each transaction in the carbon market can be recorded immutably, creating a secure ledger of emissions trades. This ensures that carbon credits are used responsibly and can't be manipulated or duplicated.

Impact: Blockchain could enhance the transparency and reliability of the carbon market, which may encourage more corporations to engage in carbon trading as a means of offsetting their emissions. A similar system could be used in a carbon card framework for tracking and rewarding sustainable behavior.

6. The United Nations Climate Chain Coalition (CCC)[84-86]

Purpose: The United Nations formed the Climate Chain Coalition (CCC), a global initiative to explore the use of blockchain technology for climate action, including tracking greenhouse gas emissions and enhancing carbon markets.

How It Works: The CCC is focused on developing blockchain-based tools to manage carbon credits, track emissions reduction, and incentivize sustainable actions. By creating an open-source platform, the CCC aims to establish a transparent and accountable system that can be adopted globally for climate actions.

Impact: The coalition's work demonstrates that blockchain has the potential to address climate challenges by improving accountability and transparency in emissions tracking and carbon credit markets, similar to the goals of a carbon card system.

Key Takeaways

These examples show that blockchain can successfully enhance transparency, traceability, and accountability in tracking emissions and incentivizing sustainability—core components of a carbon card system. The decentralized, immutable nature of blockchain helps prevent fraud, double-counting, and manipulation, which are critical for a trustworthy carbon management system.

Alternatives to Blockchain for Carbon Emission Tracking

There are several alternatives to using blockchain for tracking carbon emissions and credits. While blockchain offers transparency, decentralization, and immutability, it may not be suitable in every situation due to its complexity, energy consumption, and regulatory challenges. Various alternatives are discussed in detail in **Appendix IX – Alternatives to Blockchain for Carbon Emission Tracking**.

B. Artificial Intelligence (AI) and Machine Learning (ML) Collection and Integration

AI and ML enhance the system's accuracy and user engagement by:

- **Real-Time Emission Estimation**: AI algorithms analyze data from various sources, such as energy consumption patterns, transportation habits, and purchasing behaviors, to estimate carbon footprints dynamically.

- **Behavioral Insights**: ML models identify patterns in user behavior, providing personalized recommendations for reducing emissions. For example, the system might suggest public transportation routes instead of car travel or recommend energy-efficient appliances.

  » ML can be used to examine the emission data and cluster the data into groups using unsupervised learning algorithms — it can classify into High, Medium-High, Medium, Medium-Low or Low emission groups — providing a way to bring awareness to the individual or business for improvement.

  » Data can be fed through specialized neural network and deep learning algorithms to estimate the future emissions and credits for businesses or individuals from prior existing data. This will provide an opportunity for taking proactive actions for entities to reduce their carbon footprint based on current usage levels.

- **Dynamic Quota Adjustment**: AI can adjust carbon quotas based on evolving environmental goals, user feedback, and regional variations.

» These technologies ensure that the system adapts to user needs while continuously improving its accuracy and effectiveness. This flexibility ensures the system remains relevant and effective in promoting sustainable practices.

C. Internet of Things (IoT) for Real-Time Tracking

IoT devices play a pivotal role in tracking emissions at the source. For individuals, IoT-enabled smart meters can monitor household energy consumption, while connected vehicles can track fuel usage and emissions in real time. For corporations, IoT sensors installed in manufacturing units, logistics networks, and industrial equipment provide granular data on carbon outputs.

For instance:

- **Transportation Tracking**: IoT devices in electric vehicle (EV) charging stations or fuel stations can automatically deduct carbon credits based on consumption.

- **Industrial Emissions Monitoring**: Factories can use IoT sensors to measure emissions from production processes, integrating this data into the eCarbon Card system.

IoT's ability to automate data collection and tracking reduces errors and enhances the accuracy of emission calculations.

Core Infrastructure for Implementation

The implementation of the eCarbon Card system requires robust infrastructure capable of supporting its technological components and ensuring smooth operations.

1. Data Collection and Integration

Data is sourced from a variety of touchpoints. Advanced analytical systems must be in place to aggregate and process data from various sources. This includes:

- Utility providers supplying energy consumption data.

- Retailers integrating point-of-sale systems to track product-related emissions.

- Transportation networks sharing data on vehicle emissions and public transit usage.

2. Cloud Computing for Scalability

The eCarbon Card platform relies on cloud infrastructure to manage the vast volume of data generated by millions of users. Cloud computing ensures high availability, and secure storage, enabling the system to adapt to increasing demand without compromising performance.

3. Centralized Monitoring and Reporting

A centralized monitoring government body, such as the "Carbon Authority of India," will oversee the system's operation. The authority will manage reporting, compliance and enforcement, ensuring alignment with national climate goals.

The Carbon Authority of India: A Central Pillar for Implementing the eCarbon Card System

To ensure the effective implementation, oversight, and long-term sustainability of the **eCarbon Card System**, the establishment of the **Carbon Authority of India (CAI)** emerges as a critical requirement. This central governmental body would act as the backbone of the eCarbon Card initiative, providing the necessary regulatory framework, monitoring capabilities, and operational guidance. The CAI would not only oversee the deployment and operation of the system but also play a pivotal role in integrating it into India's broader climate policies and decarbonization goals. Its creation would demonstrate India's commitment to bold, actionable steps in its journey toward a low-carbon economy.

The Carbon Authority of India would have a multifaceted role encompassing regulation, technology management, public engagement, and policy alignment. Its responsibilities would include:

1. **Regulatory Oversight:** The CAI would establish and enforce the regulatory framework for the eCarbon Card system. This includes defining annual carbon allowances for individuals and businesses, setting sector-specific benchmarks, and creating mechanisms for penalties and rewards. The authority would ensure that allowances are equitable, adaptable, and aligned with India's decarbonization targets.

2. **Data Monitoring and Verification:** Central to the CAI's role would be the collection, analysis, and verification of carbon emissions data. Using advanced digital tools like blockchain, IoT, and AI, the authority would ensure accurate tracking of carbon transactions, preventing fraud, double counting, or system manipulation. By maintaining a robust database, the CAI would provide policymakers with real-time insights into national and regional emissions trends.

3. **Carbon Credit Marketplace Management:** The CAI would oversee the carbon trading marketplace where individuals and businesses can buy or sell carbon credits. It would establish guidelines for trading, ensure price stability, and prevent speculative market behavior. By fostering transparency and trust in the marketplace, the CAI would incentivize sustainable behaviors while creating economic opportunities.

4. **Integration with National Climate Policies:** The CAI would align the eCarbon Card system with India's broader climate commitments, such as its Nationally Determined Contributions (NDCs)[62,63] under the Paris Agreement[5]. The authority would coordinate with other government agencies, including the Ministry of Environment, Forest, and Climate Change, to ensure policy coherence and synergy across decarbonization initiatives.

5. **Public Engagement and Education:** Recognizing that public buy-in is essential for the success of the eCarbon Card, the CAI would lead awareness campaigns to educate

citizens and businesses about the system. By fostering understanding and acceptance, the authority would promote widespread participation and ensure that the system is perceived as fair, transparent, and beneficial.

6. **Adapting to Evolving Needs:** The CAI would have the flexibility to update allowances, policies, and system features in response to technological advancements, economic shifts, or feedback from stakeholders. This adaptability would ensure the long-term relevance and effectiveness of the eCarbon Card system.

The establishment of the Carbon Authority of India would have far-reaching impacts on India's decarbonization efforts.

1. **Enhanced Accountability and Transparency:** By centralizing oversight, the CAI would ensure accountability across all levels — individuals, businesses, and government bodies. Its transparent data collection and reporting systems would enable stakeholders to track progress toward emission reduction goals and hold themselves accountable for their carbon footprints.

2. **Economic Benefits through Market Efficiency:** The CAI's management of the carbon credit marketplace would create a structured, efficient trading environment that maximizes economic value while encouraging low-carbon behaviors. By setting price floors and ceilings, the authority would stabilize the market, making carbon credits a reliable economic instrument for sustainable development.

3. **Empowering Policymakers with Data-Driven Insights:** Through its advanced data monitoring and analysis capabilities, the CAI would provide policymakers with actionable insights into emission trends, sectoral performance, and regional variations. This data would enable evidence-based decision-making, helping the government design targeted interventions for high-emission sectors or regions.

4. **Strengthening Global Leadership:** By establishing the CAI, India would position itself as a global leader in climate innovation and accountability. The eCarbon Card system, supported by the CAI, could serve as a model for other countries seeking to implement equitable and scalable carbon management systems.

5. **Job Creation and Skill Development:** The operationalization of the CAI would create numerous employment opportunities, from technical roles in data analysis and system development to administrative and outreach positions. This would support India's workforce development goals while fostering a green economy.

The creation of the CAI would require a phased approach, beginning with the development of its legislative framework. A parliamentary act or executive order would define the authority's mandate, structure, and powers. Following this, the government would allocate resources for its establishment, including funding, staffing, and technological infrastructure. Partnerships with technology providers, research institutions, and international organizations would support the development of CAI's digital platforms and analytical tools. Pilot programs could be conducted to test the authority's functionality and gather feedback, ensuring a smooth nationwide rollout.

By establishing the Carbon Authority of India, the government would not only lay the foundation for the successful implementation of the eCarbon Card system but also demonstrate a bold commitment to sustainable development. This centralized body would serve as the linchpin of India's decarbonization strategy, enabling the country to achieve its climate goals while fostering accountability, innovation, and economic resilience.

High-Level Roadmap for Establishment

A phased roadmap ensures the structured and adaptive implementation of the eCarbon Card system:

1. **Feasibility Studies and Pilot Programs** Pilot programs in high-emission cities like Delhi or Mumbai, or industries such as manufacturing, will test system functionality and user engagement. Data collected during these pilots will guide refinements.

2. **Infrastructure Development** The digital platform, mobile app, and IoT network are developed and integrated. Blockchain technology is implemented to secure transactions, and the carbon credit marketplace is established.

3. **Stakeholder Engagement** Corporations, industry leaders, NGOs, and policymakers are engaged to ensure widespread support. Public awareness campaigns educate citizens about the system and its benefits.

4. **Nationwide Rollout** The system is scaled up across India, covering diverse demographics and sectors. Incentives are introduced to encourage early adoption.

5. **Continuous Monitoring and Adaptation** Regular reviews assess system performance and user feedback, enabling continuous improvement. Policies are adjusted to align with evolving climate goals.

Challenges and Strategies for Establishing the Required Technologies and Infrastructure

1. Integration with India's Existing Digital Infrastructure

Challenge:
Linking the eCarbon Card with Aadhaar (identity), UPI (payments), DigiLocker (records), ONDC (commerce), GSTN (business registry) and energy/utilities data is complex. Each system has its own architecture, data protocols, and ownership.

Strategy:
- Create secure, open APIs to allow seamless data flow between systems while preserving privacy.

- Establish a central tech coordination unit under

the Carbon Authority of India (CAI) to manage interoperability protocols.

- Use Aadhaar-based authentication for individual carbon wallets while ensuring consent-based data access.

2. Ensuring Access in Low-Connectivity Areas

Challenge:

Large parts of rural and remote India lack stable internet or smartphone access. A purely app-based solution would exclude millions.

Strategy:
- Build multi-access platforms:
  » Smartphone app for urban and digital-savvy users

  » SMS and USSD-based tools for feature phone users

  » eCarbon kiosks in panchayats, ration shops, and post offices

- Partner with BharatNet and telecom providers to expand 4G/5G and WiFi hotspots in key regions.

- Develop lightweight apps with offline functionality that sync data when online.

3. Real-Time Carbon Tracking Mechanisms

Challenge:
Accurately tracking carbon emissions from millions of daily transactions — transport, food, energy, purchases — is a data-heavy task. Most people don't have access to IoT-enabled devices or smart meters.

Strategy:
- Use approximation algorithms and activity-based emission factors for common behavior patterns (for example, petrol usage, LPG refills, electricity bills, grocery bills).

- Integrate with UPI and GSTN data to automatically categorize transactions (for example, flights, fuel, appliances) with corresponding carbon footprints.

- Provide manual entry options with AI-suggested defaults for offline or unregistered activities (for example, wood fuel, bullock carts).

- Pilot IoT smart meters in higher-emitting households and businesses to improve accuracy over time.

4. Cybersecurity and Data Privacy

Challenge:
The eCarbon Card system will handle sensitive personal and financial data, creating risks of breaches, misuse, and surveillance fears.

Strategy:

- Use blockchain or distributed ledger technologies for secure, transparent carbon credit transactions and tamper-proof records.

- Adopt privacy-by-design principles:
  - » Minimal data storage
  - » Strong encryption
  - » Consent-based access control

- Set up a Citizen Data Oversight Board under CAI to monitor data use, address grievances, and build trust.

5. Carbon Accounting for Small Businesses and Informal Sector

Challenge:

India's economy has a huge informal and MSME component, where emissions data is neither tracked nor reported systematically.

Strategy:

- Design simple carbon calculators tailored for different MSME sectors (for example, textiles, brick kilns, dairies).

- Provide incentives to digitize through carbon tracking apps and offer training via SHGs, FPOs, or cluster-based workshops.

- Use sector-specific emission averages and purchase data to estimate emissions for those not yet fully digitized.

6. Scalability and Load Management

Challenge:
A system covering over a billion individuals and tens of millions of enterprises must handle huge data volumes, real-time processing, and continuous up time.

Strategy:
- Use cloud-native architecture (e.g., AWS, NIC Cloud, Bharat Cloud) with horizontal scalability.

- Build distributed data centers and edge computing nodes across regions to reduce latency.

- Implement AI-driven load balancing and predictive maintenance for system reliability.

7. AI and Analytics for Personalization and Nudges

Challenge:
To be effective, the system must not just track carbon but also influence behavior through actionable insights.

Strategy:
- Develop AI models to analyze user behavior, suggest carbon-saving alternatives, and provide real-time feedback.

- Introduce personalized dashboards, emission trends, and gamified milestones.

- Enable "nudges" based on behavioral economics — like time-bound challenges, peer comparison, and recognition (for example, "Green Star of Your Ward").

8. Financing and Public-Private Partnerships

Challenge:
The upfront cost of this infrastructure is large and recurring operational costs (maintenance, data handling, updates) could be burdensome for the government.

Strategy:
- Mobilize green finance, climate tech funds, CSR allocations, and blended capital.

- Create a Digital Public Infrastructure (DPI) consortium involving IT firms, banks, telcos, and carbon offset platforms.

- Leverage international funding from institutions like the World Bank, GCF, UNDP, and climate-focused VC funds.

Conclusion: Building a Future-Ready Infrastructure

By integrating advanced technologies such as blockchain, AI, and IoT, and supported by a robust infrastructure, the eCarbon Card system offers a transparent, secure, and inclusive platform for managing carbon emissions. Establishing the technological backbone of the eCarbon Card system is not just an engineering challenge — it's a governance and inclusion challenge. It requires thoughtful integration with India's digital public infrastructure, strategic partnerships, phased deployment, and deep respect for the social context. If designed well, it can serve as a global template for how digital tools can democratize climate action.

Chapter 4

Mechanism for Determination of Carbon Credit Pricing

Creating a Fair, Functional, and Forward-Looking Carbon Market

Why Pricing Carbon Matters

At the heart of the eCarbon Card system lies one of the most powerful levers in climate economics: the price of carbon. A well-designed carbon pricing mechanism is not just an accounting tool — it is a market signal, a market regulator, a social equalizer, and a climate compass for a carbon-constrained world. The price of carbon credits exchanged between individuals with excess and those with a deficit can be determined through a combination of market dynamics (supply and demand), government regulations (interventions), and consideration of global benchmarks.

This chapter unpacks how carbon credit pricing could be determined in a way that is equitable, efficient, adaptive, and context-specific.

The Foundations of a People-Centric Carbon Market

In its simplest form, carbon credits are units of emissions — either allocated or saved —that can be traded. For individuals, small businesses, or farmers using the eCarbon Card, these credits become a new form of climate currency.

The value of this currency is shaped by:

- Market demand and supply
- Government policy interventions
- Global benchmarks and future goals

Together, these elements form the architecture of a resilient and responsive pricing mechanism.

A. Market-Driven Pricing: Letting Supply and Demand Set the Pace

A core pillar of pricing is supply and demand via digital peer-to-peer marketplace. As more people emit beyond their allocated allowances, the demand for credits rises, pushing prices up. Conversely, when many individuals save and trade surplus credits, supply increases, and prices fall.

This dynamic system allows for:

- Flexibility during peak seasons (e.g., festivals or summer months)

- Responsiveness to regional or lifestyle differences

- Incentives for behavioral shifts (e.g., carpooling, energy efficiency)

Such a system is inherently self-correcting, but only if:

- Participants have access to real-time price data
- Trades are conducted with low friction
- Market abuse is tightly monitored

Over time, the carbon credit market will settle into an equilibrium that reflects real-world choices and constraints — rewarding those who emit less and nudging those who emit more.

B. Government Safeguards and Policy Interventions: Setting Floors, Ceilings, and Guardrails

To prevent volatility and ensure affordability, the government must intervene in targeted ways. A **price floor** ensures that credits don't become so cheap that they lose impact or incentivizing power, while a **price ceiling** prevents costs from becoming punitive and that the market remains inclusive and protects low-income participants.

For instance, a price floor of ₹100 (1 USD = ₹86) per ton of CO_2 ensures that sellers—often low-emitting rural citizens — receive fair compensation. A ceiling of ₹1,000 per ton prevents middle-income users from being priced out when they occasionally exceed their limits. These measures keep the market stable and inclusive.

As carbon reduction goals tighten, the government may gradually increase the price of carbon credits to reflect higher targets ensuring alignment with India's climate objectives.

The government can also use rolling averages of recent trades to smooth out spikes and dips, publishing real-time prices via the eCarbon platform to foster transparency and trust. This would provide a stable reference for participants and discourage short-term arbitrage.

1. Tiered and Dynamic Pricing: Behavior-Based Incentives | A Behaviorally Intelligent Architecture

Carbon pricing should reflect not just how much one emits, but why and how. A tiered pricing structure for government-issued carbon credits could involve setting progressively higher prices as individuals purchase more credits to cover their deficit. This structure encourages responsible usage and discourages excessive reliance on government credits, maintaining incentives for sustainable behavior. This structure discourages excessive emissions while supporting flexibility for occasional overages and allows differentiation between small and large deficits, and between low-income and high-income users.

Table 4.1

Tier	Usage Above Allowance	Price (INR/Ton)	Purpose
Tier 1	0–10%	₹200	Gentle nudge, low deterrent
Tier 2	10–30%	₹600–1,000	Significant signal to curb overuse
Tier 3	30%+	₹1,200–2,000	High deterrent, luxury emissions
Emergency	Crisis/ emergency only	₹1,500+	For verified emergencies only

(1 USD = ₹86)

This structure is designed to encourage responsible usage, while recognizing that emergencies or temporary overages happen. It also supports progressive pricing — those with greater means and higher emissions pay more to reflect environmental opportunity costs, creating both environmental and social justice.

2. Government-Issued Credits: The Backstop for Market Gaps | Premiums as Penalties

Markets can fail. In times of excessive demand or supply shortages, individuals may not find credits at reasonable prices. In such cases, the government must act as a market stabilizer by issuing additional credits at a premium.

These government-issued credits are:

- Limited in quantity to preserve the market's integrity

- Priced higher than average to discourage routine use

- Tiered to reflect increasing costs with increasing deficits

- Capped per individual to avoid dependence

- Issued only after verification that an individual did indeed attempt to purchase credits from the marketplace first.

For example, if market credits average ₹500 (1 USD = ₹86) per ton, government-issued ones might cost:

- ₹700 for Tier 1
- ₹900 for Tier 2
- ₹1,200 for Tier 3
- ₹1,500 for emergency purchases

The penalty pricing model ensures that these credits are available as a safety net — but not as a first resort. This reinforces the principle that sustainability is the preferred path, not just the cheaper one.

There would be limitations or a cap on the amount of credits an Individual can purchase annually to encourage responsible usage and prevent overconsumption. For example, an individual might be allowed to purchase only 20-30% of their total allowance from the government, with the rest needing to come from personal reductions or the open market.

The influence of seasonal and regional variations would need to be considered –

- **Seasonal Adjustments**: Prices may vary seasonally due to changes in demand, with higher prices during high-consumption periods (e.g., winter, holidays) and lower prices in low-demand times.

- **Regional Pricing Tiers**: In areas with limited access to clean energy or higher fuel demands, credit prices could be adjusted to account for local conditions, ensuring fair access across different regions.

For unforeseen high-demand situations (e.g., emergencies, extreme weather), the government could create an "emergency pool" of credits available at a higher price to avoid economic hardship for individuals who genuinely need more credits.

A periodic review and adjustment — via monitoring market conditions and adjusting the availability and pricing of government-issued credits as needed — will ensure that the market remains balanced and aligned with national emissions reduction goals.

Following are various scenarios or use cases as an example –

Scenario 1: Urban Commuter

An urban user exceeds their carbon allowance due to frequent private car usage.

- They enter Tier 2.
- Credit price: ₹900/ton CO_2e (1 USD = ₹86)
- Choices:
  » Buy credits from rural users at that rate
  » Switch to metro next month and earn bonus credits

Outcome: Cost signal + Choice = Behavioral Shift

Scenario 2: Farmer with Surplus Credits

A rural farmer using biogas and shared transport emits only 60% of their annual allowance.

- Surplus: 2 tons
- Marketplace rate: ₹600/ton
- Revenue: ₹1,200 credited to e-wallet, redeemable for fertilizer or LPG

Outcome: Low-carbon lifestyles are economically rewarded

Scenario 3: Small Business in Tier 3

A small textile business exceeds its Scope 2 emissions due to inefficient energy usage.

- Over by 35%
- Tier 3 price applies: ₹1,200/ton
- Incentivized to:
  - » Buy credits (expensive)
  - » Shift to solar (cheaper over time)
  - » Earn offsets through waste recycling

Outcome: High-cost triggers innovation

3. Revenue Generation and Reinvestment in Green Projects

The carbon credit generates a predictable revenue stream that can fund green infrastructure and low carbon initiatives across India. For example,

- Renewable energy subsidies

- Clean cooking programs

- Rural electrification or clean energy projects

- Solar mini-grids

- Funding Carbon Offset programs — reforestation, renewable energy in rural areas further balancing the carbon credit system and promoting long-term, emissions reductions

- Funding climate adaptation funds

This creates a virtuous cycle, where those who exceed limits help fund sustainable infrastructure that lowers emissions — especially the poor. A transparent portal could track how much revenue is raised and where it is spent, turning carbon pricing into a visible mechanism of national development.

4. Real-Time Auctions and Dynamic Pricing Tools

To ensure efficient price discovery, the government can run initial auctions of credits at the start of each quarter or season. Bidders set prices based on perceived demand, establishing a reference point for the open market. Thereafter, prices can fluctuate based on trades, anchored by the auction baseline and bounded by the floor/ceiling range. This model — already used in power, telecom, and emissions markets globally — brings discipline and flexibility. This professionalizes the marketplace and integrates with the broader green finance ecosystem.

5. Credit Differentiation: Not All Carbon is Equal | Impact-Based Premiums

Not all carbon savings are equal. To reflect not just volume but quality of emission reductions, the system can apply pricing premiums to credits that generate co-benefits:

- Women-led community offsets: +15%
- Tree planting /Afforestation initiatives: +10%
- Health-linked projects (for example, cookstoves): +5%

These impact-weighted credits recognize that carbon abatement can also advance health, gender equity, and rural livelihoods.

Examples from Real Life: Microeconomics in Action

Imagine a family of four in an urban area that exceeds their monthly carbon allowance by 50 kg. The market rate is ₹0.80/kg (₹800/ton). They pay ₹400 to cover the gap. Next month, they choose to take the metro, lowering emissions and saving credits they can trade.

Now picture a rural farmer who uses biogas and emits just 60% of their allowance. They sell 2 tons of surplus at ₹600/ton, earning ₹1,200 — enough to subsidize next season's seeds or solar lights.

In another case, a small factory exceeding its quota enters Tier 3, facing a price of ₹1,200/ton. It chooses to install efficient machinery, cutting emissions and earning future credits. The system nudges choices, enables rewards, and builds accountability.

This creates a "carbon-plus" economy, where emissions are tied to development outcomes and social equity.

6. Price Volatility, Abuse, and Market Design

All markets are prone to abuse — especially in times of scarcity. The eCarbon Card system can mitigate this through:

- Trade limits per week or month

- Verification of real carbon activity via IoT, smart meters, or apps

- Credits Token expiration dates or storage limits to prevent hoarding

- Blockchain registries for transparent audits

These tools transform the pricing system into a trustable, auditable public utility, not a speculative marketplace.

7. Sector-Specific Dynamics and Considerations

High-emission sectors like construction, logistics, and manufacturing may need tailored pricing models that account for:

- Scope 1–3 emissions
- Efficiency benchmarks
- Regional energy availability
- Carbon leakage risks

These sectors may face:
- Carbon budgets that decline over time
- Subsidies for innovation
- Mandatory carbon disclosure linked to pricing tiers

For MSMEs and corporations, credit pricing has additional layers:
- **Forward contracts**: Pre-buy credits in bulk to hedge prices
- **Project-based offsets**: Use surplus to sponsor sectoral green upgrades
- **Carbon taxes vs carbon credits**: Pricing ensures

that credits are cheaper than government penalties, keeping incentives aligned

Example:
- Logistics company sees fuel efficiency fall → enters Tier 3
- Market price: ₹1,200/ton
- Cost of route optimization software: ₹40,000
- Emission savings: 20 tons/year
- ROI in less than a year → drives clean innovation

This ensures that businesses have both carrots and sticks aligned with their emissions profile and capabilities.

Governance and Oversight: Who Regulates the Price

While peer-to-peer market mechanisms will largely set price, an independent regulatory authority — such as the proposed **Carbon Authority of India (CAI)** — must:

- Monitor and publish daily average prices

- Intervene during price shocks (e.g., climate emergencies)

- Approve "credit class" structures and scoring algorithms

- Audit to prevent price gaming, hoarding, or manipulation

Pricing systems must be trusted. A proposed **Carbon Authority of India (CAI)** would:

1. Monitor and publish average prices and market indicators
2. Certify and verify credit quality
3. Ensure transparent auctions
4. Conduct audits to prevent price gaming, speculative hoarding, or manipulation
    *   Intervene during price shocks - pandemics, heatwaves, fuel shortages — the CAI could release emergency credits, pause price increases, or introduce temporary subsidies. This makes the system resilient and responsive.

5. Adjust pricing bands based on climate targets

C. Benchmarking Against Established Carbon Markets | Linking with Global Carbon Markets

Carbon prices in systems like the EU ETS ($80–$100/ton) and voluntary markets (Verra, Gold Standard: $5–$15/ton) offer reference points. India's pricing may start lower — ₹300–₹750/ton to account for purchasing power (parity) and to fit India's context – balancing the average individual's financial capacity with the need for meaningful emissions reduction.

As the eCarbon Card matures, carbon credit price parity with global voluntary or regulated markets is expected to evolve and should begin to align with international carbon markets, allowing high-quality Indian credits to be traded abroad, and enabling corporations to offset emissions globally.

Linking with blockchain-based platforms could further enable interoperable carbon trading, placing Indian communities and innovators into the global decarbonization economy. This allows surplus Indian credits — especially community or nature-based — to become tradable carbon assets, generating foreign climate capital for rural India.

Conclusion: Carbon Price as Compass, Not Just Currency

Pricing carbon is not about creating a cost — it's about assigning value to invisible externalities. The eCarbon Card system makes this visible, personal, and dynamic.

Done right, the carbon credit price becomes:

- A climate compass that guides behavior

- A social equity tool that redistributes value

- A market signal that supports green innovation

- A national narrative that tells citizens their actions matter

This chapter has outlined a model that is flexible, fair, and firmly rooted in behavioral insight. It combines global market logic with local realities, economic efficiency with environmental urgency, and microeconomic detail with macro-level justice. Allowing individuals with deficits to purchase credits from the government ensures that the carbon credit system remains functional and fair, even during times of high demand. By carefully controlling the availability, price, and caps on government-issued credits, India can maintain a balanced and effective carbon market that aligns with national climate goals, supports green investments, and encourages sustainable behavior across the population. This pricing model will require iteration. It will need to be calibrated, challenged, adjusted. But its very architecture — adaptive, inclusive, behavioral — ensures that it grows with the people and the planet it serves.

When set correctly, it tells users: "Your choices matter, your actions are valued, and your savings are shared".

Chapter 5

Roadmap for Implementing the
eCarbon Card System

Implementing the eCarbon Card system is a multi-faceted endeavor requiring meticulous planning, stakeholder collaboration, and advanced technological frameworks. This chapter outlines a phased roadmap designed to guide the rollout of the system, addressing both individual and business participation while ensuring adaptability and inclusivity.

Phase-Wise Implementation Plan

The implementation of the eCarbon Card system is envisioned as a three-phase process. The first phase focuses on pilot programs, the second on expansion to major urban and industrial centers, and the third on nationwide adoption with potential integration into global carbon markets.

Phase 1: Pilot Programs and Regional Rollout (Year 1)

The initial phase begins with the selection of high-emission areas and industries for targeted pilot programs. Major metropolitan areas like Delhi, Mumbai, Bengaluru, Chennai and Kolkata and industrial hubs with significant emissions would be ideal testing grounds. These pilot programs will allow the collection of real-world data to refine the system's allowances, tracking mechanisms, and user interface. Metrics will be set up to track and assess the effectiveness of the pilot program. Participants will provide valuable feedback, enabling adjustments to ensure the system's effectiveness and user-friendliness.

Additionally, pilot programs serve as an opportunity to test the robustness of technological infrastructure. IoT sensors, blockchain integrations, and mobile applications will undergo rigorous testing to identify potential issues and optimize performance. These pilots will also include education campaigns to familiarize users with the concept of carbon allowances and credits, ensuring informed participation.

Phase 2: Expansion to Major Urban and Industrial Centers (Year 2)

Building on the insights gained from pilot programs, the system will expand to second tier cities — Hyderabad, Pune, Jaipur, Indore, Ranchi, Ahmedabad, Vadodara, Chandigarh, Kanpur, Lucknow, Kochi, Coimbatore, Guwahati etc. cities and high-emission industries, such as manufacturing, transportation, and retail. This phase includes scaling the infrastructure required for the system's operations, integrating the system with existing digital tools like Aadhar, PAN Cards and UPI and launching widespread awareness campaigns to educate businesses and individuals. Training programs and workshops will be crucial to encourage adoption and promote understanding of the system's benefits.

The expansion will also involve partnerships with local governments, NGOs, and private sector players to ensure smooth implementation. For instance, corporations can play a pivotal role in promoting adoption by integrating carbon tracking into their supply chains and incentivizing their employees to participate in the system. Especially, companies committed to ESG (Environmental, Social, and Governance) can play a huge role by encouraging and helping facilitate its adoption by their employees. Coordination with municipal authorities will ensure that public infrastructure, such as transportation and utilities, aligns with the system's objectives.

Phase 3: Nationwide Rollout and Integration with International Carbon Markets (Year 3)

The final phase involves the nationwide rollout of the eCarbon Card, making it accessible to all citizens and businesses across India. This phase includes linking the system with international carbon markets, enabling participants to trade credits globally. Such integration will provide opportunities for India to participate in global carbon credit trading while aligning with international climate goals.

At this stage, the system will also focus on refining the carbon credit marketplace, ensuring that credit valuations are fair, transparent, and reflective of market dynamics. Public reporting mechanisms will be enhanced to build trust and showcase the system's contribution to national and global decarbonization efforts. Continuous stakeholder engagement will remain a priority to address challenges and gather feedback for ongoing improvements.

Government programs and initiatives such as *Swachh Bharat Abhiyan, Namani Gange* Program (Clean Ganga), or the

Ministry of Environment, Forest and Climate Change (MoEFCC) programs can be leveraged to support aligned projects.

Implementation Strategies for Individuals

For individuals, the success of the eCarbon Card system depends on transparent carbon allowance allocation, effective tracking mechanisms, and meaningful incentives.

To begin, each citizen will be allocated a personal carbon allowance based on demographic factors, such as location, household size, and lifestyle. A robust digital interface, accessible through mobile applications, will enable individuals to track their daily carbon usage. This tracking will encompass activities such as energy consumption, travel, and purchases. The app will provide insights and recommendations to encourage low-carbon choices.

Incentives will play a vital role in motivating behavior change. Individuals who stay within their annual allowances will be rewarded with discounts, tax rebates, or additional carbon credits. Conversely, those who exceed their quotas may face penalties, such as higher costs for government-issued credits. Public engagement campaigns, workshops, and gamified features within the app will enhance user participation and understanding.

To further incentivize participation, the app could integrate with popular digital payment systems, making it easier for users to redeem rewards or purchase additional credits. Personalized dashboards will display a user's carbon footprint trends, offering tips for improvement and highlighting milestones achieved in reducing emissions. This personalized approach fosters a sense of ownership and pride in contributing to national decarbonization goals.

Implementation Strategies for Business Entities

The implementation strategy for businesses involves setting baseline emissions and providing carbon allowances aligned with industry benchmarks. Businesses will be required to track their emissions across operations, including production, logistics, and supply chains. Real-time tracking systems integrated with IoT devices and blockchain technology will ensure transparency and accountability.

To facilitate compliance and incentivize reductions, businesses exceeding their allowances will have the option to purchase additional credits through a trading platform. This system will encourage investment in carbon offset projects, such as reforestation or renewable energy initiatives. Annual audits and compliance checks will ensure adherence to the system's regulations, with penalties for non-compliance.

The eCarbon Card regulatory body or implementing authority — Carbon Authority of India will actively encourage businesses to adopt innovative practices to reduce emissions through incentives, guidance, and industry engagement. For example, manufacturing units can invest in energy-efficient machinery, while logistics companies can optimize routes and adopt electric or hybrid vehicles. Sector-specific workshops will be organized to share best practices and successful case studies, fostering a collaborative approach to achieving emissions reductions.

Implementation Across the Business or Manufacturing Operation Value Chain

The concept of a Carbon Card can be integrated into the entire value chain of material production, from raw materials to finished products. The goal would be to create a robust tracking, reporting, and incentivization system for carbon emissions at every stage. This approach would provide a comprehensive overview of carbon emissions throughout the lifecycle of a product and encourage carbon-conscious decision-making among all stakeholders. Detailed roadmap for implementation is given in **Appendix X– Step-by-Step Roadmap for Implementing a Carbon Card System Across the Value Chain**.

Key Benefits of a Carbon Card System Across the Value Chain

1. **Transparency and Accountability**: By assigning a carbon footprint to each step, participants are more aware of their impact, making the entire supply chain more transparent.

2. **Incentivizing Sustainability**: Through rewards and penalties, all entities are motivated to adopt greener practices.

3. **Consumer Awareness**: Consumers can make informed choices about the carbon impact of products, fostering a market for sustainable goods.

4. **Circular Economy Encouragement**: By rewarding recycling and responsible disposal, carbon cards encourage circular economy practices.

5. **Reduced Carbon Emissions**: The entire value chain shifts towards low-carbon practices, resulting in significant emissions reductions.

Key Technological Requirements

The eCarbon Card system relies on a robust technological foundation. Blockchain technology will secure and transparently track carbon credit transactions, preventing fraud and ensuring data integrity. IoT-enabled smart meters and sensors will provide real-time monitoring of emissions, enabling accurate deductions from carbon allowances. A centralized cloud platform will manage the system's vast user data, while machine learning algorithms will refine carbon calculations based on user behaviors and regional factors.

Mobile applications will serve as the primary interface for users, offering features such as daily usage tracking, reward redemption, and carbon-saving tips. API integrations with government databases, financial institutions, and retail partners will streamline data sharing and reward distribution.

The integration of artificial intelligence (AI) will further enhance the system's capabilities. AI-driven insights can identify patterns in carbon consumption, offering predictive analytics to help users and businesses make proactive adjustments. For instance, AI could forecast high-carbon activities based on seasonal trends and suggest alternatives to reduce emissions during peak periods.

Carbon Calculation Engine based on validated algorithms will be developed as a system that accurately calculates the carbon footprint of user activities, such as food purchases, travel, and electricity consumption for individuals and Scope 1, 2 and 3 emissions for businesses and manufacturers. This could be based on a database of average emissions by category and region. Reputable climate data firms could help create an accurate carbon footprint calculation system as well. All the carbon footprint calculations would finally be fed into a centralized database viz. National Carbon Data Hub that tracks carbon metrics for various can calculate carbon impacts of each transaction.

As a User Interface, a user-friendly app (multilingual as necessary) will be developed that visually displays a user's carbon footprint, goals, progress, remaining yearly allowance and actionable tips to reduce their footprint. The app should allow users to see the carbon impact of their transactions and provide recommendations for lowering their emissions. A Personal Carbon Wallet will be introduced with a feature where users can accumulate "green credits" and redeem them for rewards. Each user's carbon footprint data can be stored securely, allowing them to track and manage their carbon balance over time.

Anti-fraud mechanisms incorporating fraud detection algorithms will need to be in place to detect unusual trading activity and prevent misuse, ensuring the integrity of the system.

Educational Campaigns and Stakeholder Engagement
Public awareness and education are essential to the system's success. National campaigns will educate citizens and businesses on the importance of carbon accountability, the mechanics of the eCarbon Card, and the benefits of participation. Workshops, webinars, and collaborations with NGOs and industry bodies will foster understanding and encourage adoption. Transparent reporting and periodic updates will build trust and demonstrate the system's impact on emissions reduction.

To ensure widespread reach, the campaigns will leverage multiple channels, including social media, television, radio, and community events. Educational materials will be tailored to different demographics, ensuring accessibility for rural and urban populations alike. Partnerships with educational institutions can further promote awareness among younger generations, embedding sustainability principles early on.

Automatic Carbon Offsetting

A small percentage of each transaction amount (for example, 1 percent) can be dedicated towards carbon offset projects, in partnership with environmental NGOs or green projects in India. This can be user-directed offset contributions by letting users choose specific projects they wish to contribute to, such as tree planting in the Western Ghats or solar electrification of villages in Rajasthan.

Continuous Monitoring and Adaptation

The eCarbon Card system will be dynamic, evolving based on user feedback, technological advancements, and progress toward national climate goals. Regular reviews of allowances, tracking mechanisms, and incentives will ensure the system's relevance and effectiveness. Emergency provisions, such as temporary adjustments to allowances during natural disasters or economic crises, will provide flexibility and resilience.

Advanced analytics and periodic audits will play a crucial role in monitoring the system's performance. Real-time dashboards for policymakers will provide insights into national and regional emissions trends, enabling data-driven decisions. Feedback loops will ensure that user suggestions and concerns are incorporated into system updates, fostering continuous improvement.

The Carbon Authority of India (CAI) - Institutional Backbone of the eCarbon Card Ecosystem

For an initiative of such unprecedented scale and transformative ambition, establishing a centralized and purpose-built governing body is not only logical — it is essential. The Carbon Authority of India (CAI) is envisioned as the regulatory, administrative, and strategic nerve center of the eCarbon Card ecosystem. Tasked with managing the complexity of a decentralized, people-centered carbon economy, the CAI would serve as the institutional anchor that ensures efficiency, equity, transparency, and adaptability in the implementation and evolution of the system.

With the eCarbon Card touching every citizen and enterprise, across every state and sector, the CAI will function as the single point of oversight and coordination, driving India's shift toward a low-carbon future. It will regulate the issuance and trading of carbon credits, set allowance benchmarks, integrate with national digital infrastructure, and maintain alignment with India's broader climate goals.

By institutionalizing the framework through which carbon accountability, behavioral incentives, and market-based mechanisms converge, the Carbon Authority of India becomes the operational backbone and moral compass of a new kind of climate governance — decentralized in execution, centralized in vision, and rooted in democratic accountability.

D. Core Mandate and Functions

At its heart, the CAI would serve as the guardian of India's carbon equity economy. Its foundational mandate includes:

- Administering and monitoring the eCarbon Card system for individuals and enterprises.

- Defining and allocating carbon allowances across demographics, regions, and sectors, based on equity, necessity, and national targets.

- Setting and enforcing rules for carbon credit trading, including transaction limits, anti-fraud mechanisms, and regulatory compliance.

- Establishing price floors and ceilings for carbon credits to prevent volatility and ensure affordability.

- Serving as a market stabilizer, stepping in during supply-demand imbalances through controlled issuance of government credits.

- Integrating the system with India's digital stack — Aadhaar, UPI, DigiLocker, ONDC — for seamless identity, payment, and data interoperability.

- Providing transparent, real-time data dashboards for policymakers, researchers, and the public.

- Running public education campaigns to drive awareness and engagement across all socioeconomic strata.

E. System Management and Oversight

The CAI would be responsible for defining and dynamically updating:

- Carbon quotas and allowances, including differentiated caps for urban vs. rural areas and essential vs. luxury activities.

- Tiered pricing policies and credit classification (e.g., nature-based, industrial, community-offset credits).

- Audit frameworks for individuals and businesses to ensure accuracy in reporting and claiming credits.

- Penalties for non-compliance, greenwashing, and credit hoarding or manipulation.

It would operate under the supervision of a multi-stakeholder council comprising representatives from:

- Central and state governments

- Climate scientists and economists

- Civil society and labor unions

- Industry associations and SMEs

- Renewable energy, technology, and data sectors

This diverse structure ensures pluralistic, decentralized feedback in a system guided by centralized standards and enforcement.

F. Data Intelligence and Adaptive Governance

One of the most powerful functions of the CAI would be real-time data analytics and dynamic policy response. By aggregating carbon transaction and emissions data from across India, the CAI could:

- Identify regional or sectoral spikes in emissions

- Detect behavioral trends in carbon consumption

- Simulate policy impacts using AI-driven modeling

- Create climate vulnerability maps for targeted interventions

- Tailor subsidies or policy nudges for high-impact areas

For example, if data shows a recurring urban transport surge in emissions during summer months, the CAI could trigger:

- Bonus credits for EV use or metro rides

- Temporary price hikes for high-carbon private travel

- Public campaigns linking heat waves to emissions

Such data-enabled governance transforms CAI from a regulatory entity into a living, adaptive institution responsive to climate signals and societal feedback.

G. Technology Stack: Securing the System of Trust

To scale with integrity, the eCarbon Card system must be tamper-proof, interoperable, and auditable. The CAI would oversee the integration of frontier technologies:

- Blockchain for secure, decentralized recording of carbon credit issuance and transactions, preventing fraud or double counting.

- IoT networks (smart meters, transport trackers, digital receipts) for automated carbon data collection.

- Artificial Intelligence to model emissions trends, forecast demand, and personalize sustainability recommendations.

- Digital Identity and Consent Frameworks (via Aadhaar/DigiLocker) to allow users to control and share their emissions data securely.

This digital architecture positions India not only as a leader in decarbonization, but also in climate-tech innovation.

H. Equitable and Inclusive Role

The CAI would also play a critical equity role in ensuring that the carbon pricing and credit system does not disproportionately burden the poor or marginalized. It would:

- Define special allowances for rural, low-income, elderly, or differently abled citizens

- Support Self-Help Groups (SHGs) and MSMEs in generating tradable carbon credits

- Offer higher allowance floors for users without access to low-carbon alternatives

- Guide public investment of carbon credit revenue into infrastructure for those most affected by climate change

Through its policies and outreach, CAI becomes a climate equity champion—making climate justice a practical reality.

I.   Integration with National and Global Frameworks
The CAI would liaise with:

- India's Ministry of Environment, Forest and Climate Change (MoEFCC) to align with national NDCs

- International climate bodies, such as UNFCCC, to ensure transparency and eligibility for global carbon markets

- Private and public carbon registries (like Verra, Gold Standard) to link India's credit system to global standards

- State governments and local panchayats to adapt
  the eCarbon Card system to regional priorities and
  infrastructure

Its long-term goal would be to position India as a global standard-setter in decentralized carbon markets, exporting high-quality credits while maintaining domestic resilience and equity.

J. Building Public Trust and Engagement

Perhaps the most important role of the CAI is to foster public confidence. A system that touches every life — from farmers to factory owners — must be transparent, accountable, and participatory.

The CAI would:

- Publish annual climate equity and carbon
  performance reports

- Maintain an open data dashboard tracking
  system outcomes

- Operate a grievance redressal portal for disputes

- Partner with educators, media, and civil society to
  build a carbon-literate population

Through this, the CAI becomes not just a regulator but a trusted custodian of India's climate future.

K. The Institutional Engine of a New Climate Compact

The Carbon Authority of India would serve as the institutional engine of the eCarbon Card system — ensuring that it is not merely a tool of personal accountability, but a nationwide contract for a sustainable, just, and resilient future.

It combines:

- Strategic oversight
- Technological integration
- Behavioral economics
- Democratic accountability
- Global connectivity

With the CAI in place, the eCarbon Card becomes not just a visionary idea, but a credible, scalable, and sovereign instrument of transformation. As climate becomes the defining challenge of the 21st century, the Carbon Authority of India ensures that every citizen, every action, and every credit counts.

Why a Phased Implementation Plan is Critical

Using a phased implementation plan as a learning engine for deploying the eCarbon Card system is not just practical — it is essential for ensuring its success, scalability, inclusivity, and long-term public acceptance. Rather than attempting an immediate nationwide rollout, a stepwise deployment allows the system to evolve through evidence-based iteration and localized adaptation.

A phased approach enables the government and implementation partners to:

1. **Start with manageable populations**
   By launching pilot programs in select districts or cities, authorities can test the system under controlled, real-world conditions, ensuring that diverse socio-economic, urban-rural, and linguistic factors are addressed early on.

2. **Test integration with India's digital infrastructure**
   The eCarbon Card system must seamlessly interface with platforms like Aadhaar (identity), UPI (payments), DigiLocker (documents), and ONDC (commerce). Pilot regions can help validate these integrations before national scale-up.

3. **Identify and resolve operational bottlenecks**
   Early phases provide critical insight into technical bugs, user interface challenges, data privacy concerns, and behavioral resistance, allowing for timely course corrections.

4. **Refine policy and pricing structures**
   Real-world pilots offer the opportunity to test and calibrate carbon allowance levels, credit pricing mechanisms, and tiered incentives in ways that are responsive to local consumption patterns and economic realities.

5. **Build community engagement and trust**
   A phased rollout supports grassroots partnerships with Self-Help Groups (SHGs), Panchayats, schools, local NGOs, and digital literacy missions — building trust, awareness, and a sense of ownership among participants.

6. Generate success stories and data

Demonstrating visible benefits — such as income generation for rural low emitters or behavioral shifts in urban areas — creates positive feedback loops that drive broader interest and participation.

7. De-risk national rollout

By addressing critical risks at a smaller scale — such as system overload, inaccurate emissions tracking, or market imbalances — India can de-risk future investment, reduce administrative strain, and build institutional confidence.

Starting with a few high-impact, high-visibility districts or sectors — such as major metros, logistics hubs, or industrial clusters — allows India to:

- Showcase the model globally as a proof of concept

- Demonstrate measurable impact quickly (carbon savings, green jobs)

- Position itself as a first mover in decentralized climate markets

What Can Be Learned Through Pilots?

Pilot regions become living labs to:

- Calibrate pricing models, communication strategies, and technological tools

- Test equity safeguards and regional customization

- Validate blockchain, AI, and IoT-based upgrades

- Stress-test data privacy, market fairness, and supply-demand dynamics

Strategic Benefits of Phasing

A phased rollout ensures:

- Infrastructure readiness before full-scale stress

- Widespread stakeholder engagement without overload

- Policy harmonization with existing government schemes (e.g., MNRE, PM-KUSUM, *Ujjwala*, PDS)

- Adaptability to emerging tech and citizen feedback

Global Relevance and Investment Appeal

Each successful pilot becomes a global case study — attracting attention from:

- Multilateral institutions

- Impact investors and green funds

- Governments seeking replicable climate action models

Conclusion

Implementing the eCarbon Card system across India's entire value chain — from raw material extraction and manufacturing to logistics, retail, and consumer use — offers a transformative pathway to embed sustainability into everyday economic life. By tracking and incentivizing emissions reduction at every stage, the system becomes more than a monitoring tool — it becomes a climate catalyst.

Such a mechanism could significantly accelerate progress toward India's net-zero target by 2070, while unlocking new avenues for green innovation, jobs, equity, and resilience. Yet for a solution of this scale to succeed, a phased, evidence-based rollout strategy is not just advisable — it is essential.

In short, a phased implementation plan transforms the eCarbon Card system into a dynamic policy innovation platform — where learning, adaptation, and participatory governance are embedded from day one. It ensures that India's decarbonization journey is not only data-driven and technologically sound, but also socially inclusive, politically feasible, and globally inspirational.

With the right sequencing, partnerships, and feedback loops, the eCarbon Card system can become a lighthouse model — not just for India, but for every nation seeking to align economic growth with ecological responsibility.

Chapter 6

Real-World Applications, Global Precedents and Case Studies

The implementation of the eCarbon Card system offers the promise of transforming carbon accountability and reduction into an accessible, actionable framework. Its practical utility can be best understood through real-world examples and case studies that showcase its application across households, businesses, and governmental programs. By drawing parallels with existing systems globally, we can better appreciate how the eCarbon Card can achieve measurable impacts in reducing carbon footprints.

Carbon Tracking in Daily Life

Incorporating carbon tracking into daily life involves a systemic change in how individuals, families, and businesses interact with their environment. Households can use the eCarbon Card to monitor energy consumption, transportation habits, and shopping patterns. For example, a middle-class family in an urban Indian setting could use the card to track emissions from their electricity usage, daily commute, and grocery purchases. The insights provided by the card would help them identify high-carbon activities, such as using private vehicles for short trips, and encourage them to opt for sustainable alternatives like public transportation or carpooling. Over time, such behavioral shifts, incentivized by the eCarbon Card's rewards system, can lead to significant reductions in household emissions.

Small businesses, too, can leverage the system to optimize operations. A local grocery store, for instance, could monitor its energy use and supply chain emissions, identifying areas to switch to renewable energy sources or adopt local sourcing to minimize transportation-related emissions. Large corporations, on the other hand, may find the eCarbon Card indispensable for achieving compliance with corporate social responsibility goals. A multinational manufacturing firm could integrate the card with its production systems to monitor emissions across different facilities, ensuring adherence to emission caps and leveraging the carbon trading marketplace to offset unavoidable emissions.

One compelling example could be that of a suburban family participating in an eCarbon Card pilot program in Bengaluru. The family initially had a monthly carbon footprint of 1.8 tons due to their reliance on private cars, high electricity consumption from air conditioners, and frequent purchases of packaged goods. Within six months of using the eCarbon Card, they reduced their emissions by 25 percent, primarily by switching to public transport, installing solar panels for their home, and opting for local farmers' markets for groceries. This case study illustrates the eCarbon Card's potential to drive tangible changes in lifestyle and consumption patterns. Furthermore, the use of gamification[31] in the app — such as setting monthly targets with rewards for achievements — made the process engaging and enjoyable, leading to sustained participation.

Corporate Responsibility and Supply Chain Optimization

Corporations worldwide are increasingly recognizing the importance of carbon accountability within their supply chains. The eCarbon Card system takes this a step further by providing a transparent framework to track and reduce emissions across complex supply chains. For instance, a textile manufacturing company could use the card to assess emissions at every stage, from sourcing raw materials to distributing finished products. By partnering with suppliers committed to sustainability, the company could not only reduce its carbon costs but also enhance its brand image as an environmentally responsible enterprise.

One example is Mahindra & Mahindra's commitment to becoming a carbon-neutral organization by 2040. By adopting the eCarbon Card system, the company could streamline its monitoring of emissions across operations, from vehicle manufacturing to farm equipment production[87,88]. Similarly, Tata Steel, with its ambitious sustainability goals to realize Net-Zero emissions by 2045, could utilize the card to quantify emissions from raw material procurement and energy use, enabling a more granular approach to meeting their decarbonization targets[89].

Unilever's initiative to source 100 percent renewable energy for its operations by 2030 provides a global parallel[90]. Through transparent reporting and collaboration with renewable energy providers, Unilever achieved significant emissions reductions while maintaining profitability. Also noteworthy is its commitment to achieving net-zero emissions across its value chain by 2039. Through strategic collaborations with suppliers and the adoption of advanced emission-tracking technologies, the company has set a benchmark for sustainable supply chain practices. Similarly, Apple's initiative to transition its entire supply chain and products to be 100 percent carbon neutral by 2030 demonstrates how corporations can lead by example[91]. The eCarbon Card system, providing granular data on emissions, could serve as a critical tool for such initiatives, enabling corporations to identify inefficiencies and drive targeted reductions. The eCarbon Card could augment such efforts by offering real-time tracking and incentivizing further reductions through carbon credit trading. Additionally, smaller enterprises could benefit from pooling resources to access sustainable technologies, with the eCarbon Card serving as a platform to incentivize collaborative projects.

Integration with Government Initiatives

The eCarbon Card system aligns seamlessly with India's existing environmental and developmental goals. Programs such as the *Swachh Bharat Mission*, the Smart Cities Mission, and the National Solar Mission are all aimed at fostering sustainability and reducing the country's carbon footprint. The eCarbon Card can complement these initiatives by introducing an additional layer of accountability and incentivization. For example, urban residents participating in the Smart Cities Mission could use the card to monitor emissions from daily activities and access rewards for adopting eco-friendly practices, such as cycling or using electric buses.

The system could also integrate with the *Ujjwala Yojana*, which promotes cleaner cooking fuels for rural households. By tracking the transition from biomass to LPG or electric stoves, the eCarbon Card could provide quantifiable data on emissions reductions, while incentivizing further adoption through rewards. In rural areas, the *Pradhan Mantri Sahaj Bijli Har*

Ghar Yojana (*Saubhagya* Scheme) could be augmented by enabling households to monitor their electricity use and optimize it for maximum efficiency, reducing emissions without compromising on energy access. These integrations not only create synergies with existing programs but also amplify their impact by providing measurable outcomes and could accelerate the adoption of clean energy across sectors.

For instance, consider a pilot program in a rural district where households using the eCarbon Card are provided additional rewards for switching to solar-powered irrigation pumps. The data collected through the card demonstrates a 40 percent reduction in emissions from traditional diesel pumps, highlighting the scalability of such initiatives across other districts. The government's ability to use this data for policymaking further underscores the value of integrating the eCarbon Card with national schemes.

The card's potential to support India's commitments under international climate agreements, such as the Paris Accord, cannot be overstated. By enabling the precise tracking of emissions at individual and corporate levels, the system provides a quantifiable method to demonstrate progress toward national targets.

Global Examples & Case Studies Relevant to eCarbonCard

While the eCarbon Card system is a novel and integrated approach combining personal carbon tracking, behavioral incentives, carbon credit trading, and national digital infrastructure, several global programs, pilots, and policy experiments echo parts of its vision. Below is a curated set of detailed global examples and real-world case studies that align with different elements of the eCarbon Card system.

1. European Union Emissions Trading System (EU ETS)[28]

Location: European Union (27 member states + Iceland, Liechtenstein, Norway)

Key Concept: Cap-and-trade system for industrial emissions

Details:

- Launched in 2005, the EU ETS is the world's largest carbon market, covering power plants, manufacturing, and intra-European aviation.

- Companies receive or purchase EU Carbon Allowances (EUAs)[92] which they must surrender based on actual emissions.

- If emissions are below allowance, the surplus can be sold. If above, more credits must be purchased.

- The cap shrinks annually, tightening emissions targets over time.

- Between 2005 and 2020, emissions from sectors under the ETS declined by more than 35 percent.

Relevance to eCarbon Card:

- The EU ETS shows that market-based carbon pricing can work at scale when properly regulated.

- It offers a mature example of dynamic credit pricing, auctioning, and emissions monitoring — principles the eCarbon Card applies to at the individual and small business level.

- The eCarbon Card builds upon the EU ETS by extending cap-and-trade logic beyond corporations to citizens, local enterprises, and informal sectors — democratizing carbon accountability.

Key Lesson for India:

- The EU ETS demonstrates that emissions trading can create an efficient, self-correcting market.

- It also reveals the importance of strong oversight, verified data, and clear penalties to maintain market integrity.

- India's Carbon Authority (CAI) could adopt a similar role to the European Commission in ensuring regulatory consistency, market fairness, and cross-sectoral alignment.

A notable international precedent is the European Union's Emissions Trading System (ETS). The ETS demonstrates the effectiveness of cap-and-trade mechanisms in reducing industrial emissions across member states. The eCarbon Card system builds upon this model by extending its reach to individuals and small businesses, creating a more inclusive and comprehensive framework for carbon management. Furthermore, the success of California's cap-and-trade program, which reinvests revenues into sustainable infrastructure, provides a roadmap for how funds generated through carbon trading can be utilized effectively.

2. UK Personal Carbon Allowance Trials (RCEP Proposal, 2006–2011)[36, 93-98]

Location: United Kingdom
Key Concept: Personal carbon budgeting
Details:

- The UK's Royal Commission on Environmental Pollution (RCEP) proposed a Personal Carbon Trading (PCT) system, where every adult would receive a carbon allowance for home energy, fuel, and travel.

- Surplus credits could be traded; individuals who needed more had to buy additional units.

- Multiple feasibility studies (DEFRA) concluded the system was technically possible, but political will and public readiness were barriers.

- A pilot by the RSA (Royal Society of Arts) in 2008 used voluntary carbon diaries, showing that participants reduced carbon emissions by up to 30% through increased awareness.

Relevance to eCarbon Card:

- Early vision of personal allowances
- Lessons in behavioral impact, tech design, and political narratives
- Challenges in scalability without integrated national systems like Aadhaar/UPI

3. UK's Emissions Trading System (UK ETS)[99,100]

Location: United Kingdom

Key Concept: A cap-and-trade system launched in 2021 to replace the UK's participation in the EU ETS post-Brexit. It sets a limit (cap) on total greenhouse gas emissions and allows businesses to buy, sell, or trade emission allowances within that cap.

Covered Sectors: Power generation, energy-intensive industries, and aviation (with future expansion planned).

Market Mechanism:

- Auctioning of allowances
- Secondary trading
- Government-set price floors and ceilings

Regulated Entities: Applies to ~1,000 UK-based businesses emitting over 20 MW of thermal capacity or 2,500 tons of CO_2/year.

Relevance to the eCarbon Card

- **Proof of Market-Based Success:**
  Demonstrates how emissions trading can effectively reduce carbon while maintaining economic

flexibility — a core principle of the *eCarbon Card* at the individual level.

- **Cap-and-Trade Blueprint:**
  Offers a robust design template for how personal and small-business carbon trading can be adapted from institutional models.

- **Policy and Pricing Precedents:**
  Provides a reference for credit auctioning, price stability mechanisms, and compliance structures — all of which can be scaled down and customized for *eCarbon Card* users.

- **Scalability Insight:**
  Highlights the importance of phased rollout, regulatory oversight, and digital infrastructure, which are critical for the successful deployment of a decentralized carbon economy in India or globally.

4. Sweden's Carbon Tax + Social Incentives (since 1991)[101-105]

Location: Sweden

Key Concept: Behavioral carbon pricing at the national level

Details:

- Sweden implemented a national carbon tax on fossil fuels in the 1990s, currently among the highest globally (~$130/ton).

- It combined this with rebates, low-income subsidies, and green investment programs.
- Result: Sweden's GDP grew by >60% since 1990 while emissions dropped by >25%.

Relevance to eCarbon Card:
- Example of price signals + social equity
- Demonstrates the feasibility of long-term carbon behavioral change at scale
- Strong citizen trust in government mechanisms is a key enabler

5. **China's Personal Carbon Footprint Tracking (Ant Forest by Alipay)[106-108]**

Location: China

Key Concept: Gamified personal carbon tracking

Details:
- Alibaba's Ant Forest, launched in 2016 via the Alipay app, tracks users' low-carbon behaviors (e.g., walking, biking, online payments instead of paper bills).

- Points earned are used to grow virtual trees, which result in real trees planted in desert areas in Inner Mongolia and Gansu.

- As of 2023, over 600 million users planted 300+ million trees, reducing CO_2 by an estimated 10+ million tons.

Relevance to eCarbon Card:
- Behavioral design using gamification, rewards, and identity
- Digital platform with huge user base across income levels
- Voluntary, but highly successful in public engagement

6. China's National Emissions Trading Scheme (China ETS)[109]

Location: China

Key Concept: Power-sector focused carbon trading scheme with phased expansion

Details:
- Launched in July 2021, China's ETS is the world's largest in terms of covered emissions, initially targeting over 2,000 power plants emitting more than 26,000 tons of CO_2 annually.

- It currently covers only Scope 1 CO_2 emissions from coal- and gas-fired power generators, accounting for more than 40% of China's total emissions.

- Unlike the EU ETS, the China ETS allocates allowances for free using a benchmarking approach based on plant efficiency rather than a hard cap.

- It is designed as an intensity-based system, meaning allowances are tied to output rather than absolute emissions caps.

- Expansion plans include extending coverage to steel, cement, aluminum, chemicals, and other high-emission sectors in coming years.

- Data reporting and verification processes are centralized under the Ministry of Ecology and Environment, with ongoing improvements to MRV (Monitoring, Reporting, Verification) infrastructure.

Relevance to the eCarbon Card:

- China's ETS shows how a phased and sector-focused approach can enable a carbon market to take root even in highly industrial and rapidly developing contexts.

- The use of intensity-based benchmarks provides an alternative to absolute caps—relevant for India's carbon allowance logic, especially in sectors where economic growth is still accelerating.

- The centralized but gradually decentralized implementation strategy offers a template for India's potential rollout through state and local agencies under a national carbon governance framework.

- The ETS highlights the importance of building MRV capacity first, before scaling up market complexity — a key lesson for India's eCarbon Card digital backbone.

Key Lesson for India:
- A sector-first, benchmark-linked design offers flexibility in economies where hard caps may be politically or developmentally infeasible.
- China's experience suggests that national carbon governance can coexist with developmental priorities — and can even reinforce energy efficiency and productivity.
- India's rollout of the eCarbon Card can emulate China's strategy of starting narrow, building infrastructure, and then broadening the scope of coverage.

7. California Cap-and-Trade System + Low-Income Rebates[29]

Location: California, USA

Key Concept: Market-based emissions reduction with equity tools

Details:
- Since 2013, California has operated a statewide carbon market for industries and power plants.
- Revenues are partly redistributed to households through California Climate Credit: every household gets a utility bill rebate funded by emissions auctions.
- Also, funds EV subsidies, transit, and energy retrofits in low-income communities.

Relevance to eCarbon Card:
- Shows how market tools + equity-based redistribution work in tandem
- Demonstrates public acceptance when climate dividends are visible

- Emissions reduced while creating green jobs and energy savings

8. Sitra's Personal Carbon Footprint App (Finland)[110]

Location: Finland

Key Concept: Real-time personal carbon footprint tracking

Details:

- Sitra, the Finnish Innovation Fund, launched lifestyle calculators and apps to help citizens track emissions from food, transport, energy, and consumption.
- Integrated with banking data and utility data to estimate carbon footprints.
- Users receive personalized recommendations to reduce emissions and set goals.

Relevance to eCarbon Card:

- Real-time tracking integrated with finance and utilities
- Data-driven behavior change with personal empowerment narrative
- Government-backed, privacy-respecting, voluntary approach

9. Estonia's Digital Governance and Carbon Data Integration[111-115]

Location: Estonia

Key Concept: Blockchain-enabled government services

Details:

- Estonia operates a fully digital government with blockchain-secured records for education, health, identity, and finance.

- It is exploring green governance tools, such as linking emissions data from utilities and transport to digital identities.
- Pilots are underway to connect energy consumption data with personal carbon accountability.

Relevance to eCarbon Card:

- Shows feasibility of a national digital ledger system
- Strong analogy for using Aadhaar + UPI + carbon wallets
- Transparent and auditable public trust architecture

10. Japan's Eco-Points Program[65] (2009–2011)

Location: Japan

Key Concept: Incentivized low-carbon consumption

Details:

- Government awarded "eco points" for purchases of energy-efficient appliances.
- Points could be redeemed for goods or services.
- Stimulated both the consumer economy and emissions reductions during the post-recession recovery.

Relevance to eCarbon Card:

- Behavior-based rewards for low-emission purchases
- Government-subsidized and verified green consumption
- Evidence of rapid mass adoption when rewards are clear and accessible

11. South Korea's Emissions Trading Scheme (K-ETS)[116-120]

Location: South Korea

Key Concept: Sectoral carbon market + real-time emissions tracking

Details:

- Covers over 600 large emitters and tracks Scope 1 & 2 emissions.
- Allows real-time emissions monitoring, third-party verification, and credit trading.
- Successful in reducing industrial emissions while enabling economic growth.

Relevance to eCarbon Card:

- Shows that digital carbon markets can function at national scale
- Model for how India could build enterprise-focused carbon allowance systems under eCarbon Card for MSMEs

12. Singapore's Carbon Tax and Green Transition Toolkit[121-124]

Location: Singapore

Key Concept: Carbon tax with enterprise innovation credits

Details:

- Introduced a carbon tax of $5/ton, rising to $50–$80 by 2030.
- Coupled with grants for SMEs and businesses to decarbonize and digitalize.

- Platforms to track corporate emissions footprints and integrate with ESG reporting.

Relevance to eCarbon Card:
- Integrated carrot and stick approach
- Applies to businesses with scalable digital frameworks
- Could inform how Indian firms report under the eCarbon Card

The table below summarizes key takeaways from these worldwide examples.

Table 6.1

Summary of Global Carbon Market Precedents and Lessons

	Country / Region	Program / System	Key Features	Relevance to eCarbon
1	United Kingdom	Personal Carbon Allowance Trials	Personal allowances & trading trials, carbon diaries	Piloted personal tracking and allowance trading model
2	Sweden	National Carbon Tax	High carbon tax +social welfare alignment	Shows pricing + equity integration at national scale
3	China	Ant Forest (Aliplay)	Gamified carbon tracking & tree planting	Behavior change via rewards and digital identity
4	China	China-ETS (National Emissions Trading Scheme)	World's largest ETS by coverage; intensity-based allocation; power sector focus	Illustrates scalable national implementation with real-time monitoring and phased rollout

	Country / Region	Program / System	Key Features	Relevance to eCarbon
4	California (USA)	Cap-and-Trade + Climate Rebates	Market based trading + direct citizen rebates	Inspiration for citizen climate dividends & policy synergy
5	Finland	Sitra Personal Carbon Apps	Carbon calculators & personal lifestyle tracking	Real-time footprint tracking model for individuals
6	Estonia	e-Governance + Carbon Tracking	Blockchain-secured digital services, emission integration	Digital infrastructure & privacy model for integration
7	Japan	Eco-Points	Rebates for low-carbon appliances & behavior	Incentivized consumer behavior, easy to scale
8	South Korea	K-ETS (Emissions Trading Scheme)	Sector-wide emissions caps, trading, monitoring	Mature cap-and-trade market structure and oversight
9	Singapore	Carbon Tax + SME Green Toolkit	Carbon tax + SME decarbonization funding	Toolkits for SMEs; scalable business-level interventions
10	European Union	EU Emissions Trading System (EU ETS)	Cap-and-trade for heavy industry, shrinking cap model	Model for national cap-and-trade adapted to individuals

Table 6.2

Key Lessons for India's eCarbon Card System

Theme	Lesson / Takeaway
Behavioral Economics	Use gamification (China), cash rebates (California), and eco-points (Japan)
Digital Integration	Estonia, Finland show digital ID and secure data linkages work at scale
Equity and Inclusion	Sweden and California embed fairness through subsidies and universal dividends
Market Design	South Korea and EU ETS show how robust markets can scale, but need oversight
Voluntary → Mandatory	Starting with voluntary systems (apps, diaries) builds trust before mandates
Global Alignment	India can position itself as a leader in decentralized carbon marketplaces

Additional Global Parallels and Insights

Globally, several initiatives mirror aspects of the eCarbon Card system, providing valuable insights into its potential implementation and impact. In Sweden, Doconomy's "DO Card" tracks the carbon footprint of users' purchases, offering a real-world example of how financial tools can promote carbon accountability. Mastercard has recently partnered with them. DBS Bank, based in Singapore, has implemented a carbon footprint tracker for credit and debit card holders[125,126]. By analyzing spending patterns, it provides users with an estimate of their carbon footprint, allowing them to understand the impact of their financial choices on the environment. Similarly, in the United Kingdom, academic studies on Personal Carbon Allowance[36-38] explore the feasibility of allocating carbon budgets to citizens, laying the groundwork for systems like the eCarbon Card.

In Australia, platforms such as Greenwallet empower individuals to monitor and reduce their carbon footprints through detailed analyses of their spending patterns[127]. These examples highlight the increasing global emphasis on personal carbon accountability. The eCarbon Card system stands out by integrating such tracking mechanisms with a robust rewards and trading framework, ensuring that users are not only aware of their emissions but also motivated to act.

Joro is a mobile app based in the United States that helps users track and reduce their carbon footprint by analyzing spending patterns[128,129]. Users can offset their carbon footprints by investing in carbon reduction projects. Similarly, Carbon-Click, based in New Zealand provides tools for individuals and businesses to track and offset their carbon footprints[130]. Although it doesn't issue a "carbon card," the idea is similar in that users can monitor their carbon impact and make adjustments to reduce their emissions. Based in the UK, CoGo is an app that connects with bank accounts to track users' carbon footprints based on spending. It gives recommendations to help users reduce their impact and suggests companies with more sustainable practices.[131]

Hypothetical India-Specific Case Studies

1. **Household Case Study: Urban Middle-Class Family**
    Consider a family of four living in a metropolitan area like Mumbai. By using the eCarbon Card, the family tracks their carbon emissions across various activities, including electricity use, daily commutes, and food purchases. Over the first year, the family identifies that a significant portion of their emissions comes from running air conditioners during peak summer months. With insights from the eCarbon Card app, they invest in energy-efficient cooling systems and reduce their electricity consumption by 20 percent, earning additional carbon credits in the process. Additionally, the family begins using public transport for their daily commute, further reducing their emissions by 15 percent. This case demonstrates how the system not only drives behavior change but also creates financial incentives for sustainable practices.

2. **Corporate Case Study: Large Manufacturing Firm**
    A cement manufacturing firm with operations across India adopts the eCarbon Card system to monitor emissions from its factories and supply chain. By analyzing data from the card, the firm identifies inefficiencies in its logistics operations and transitions to using rail freight instead of road transport for raw material delivery. This shift reduces transportation-related emissions by 30 percent, enabling the company to stay within its annual allowance and sell surplus credits on the carbon trading marketplace. Furthermore, the

firm invests in alternative fuels, such as biomass and waste-derived fuels, achieving an additional 10 percent reduction in overall emissions.

3. Government Integration Case Study: Renewable Energy Initiative

In partnership with the Ministry of New and Renewable Energy, the eCarbon Card system is piloted in rural villages transitioning to solar power under the *Saubhagya* scheme. Residents use the card to track their reduced reliance on kerosene lamps, with data showing a 50 percent decrease in household emissions within the first six months. The success of the pilot leads to nationwide adoption, with the eCarbon Card becoming a standard tool for tracking progress in rural electrification programs. Moreover, the integration of micro-financing options for purchasing solar home systems ensures that the program is both accessible and impactful.

4. Urban Transportation Case Study

In Delhi, the eCarbon Card is integrated into the city's public transportation system. Commuters using metro services and electric buses receive carbon credits, which can be redeemed for reduced ticket prices or discounts on eco-friendly products. Over two years, public transport ridership increases by 15 percent, significantly reducing vehicle emissions and congestion in the city. The data collected also allows policymakers to identify high-demand routes and allocate resources more efficiently, further optimizing the city's transit network.

Conclusion

The real-world applications and case studies of the eCarbon Card system illustrate its transformative potential in promoting carbon accountability and fostering sustainable practices. By drawing on insights from global initiatives and tailoring its approach to India's unique challenges, the system can drive significant reductions in emissions across sectors. Whether through empowering individuals, optimizing corporate supply chains, or supporting government programs, the eCarbon Card system represents a powerful tool for achieving India's decarbonization goals while setting a precedent for other nations to follow.

By incorporating advanced technologies, engaging stakeholders, and leveraging incentives, the eCarbon Card system offers a compelling solution to one of the most pressing challenges of our time. Its real-world impact, as demonstrated through these examples, underscores its viability as a scalable model for sustainable development worldwide.

Chapter 7

*Economic and Environmental
Impact and Calculations*

Meeting Decarbonization Goals and Job Creation

The eCarbon Card system presents a transformative approach to reducing carbon emissions while simultaneously fostering economic growth. By meticulously calculating allowances and developing robust mechanisms for carbon credit trading, this system has the potential to significantly impact both environmental and economic spheres. This chapter explores the methodologies behind calculating carbon allowances, the projected emissions reductions, and the broader economic implications of implementing such a system in India. Moreover, examples and case studies are interwoven to highlight the practicality and effectiveness of this innovative model.

Carbon Allowance Calculation

The foundation of the eCarbon Card system lies in accurately calculating yearly carbon allowances for individuals and businesses. This process begins with setting a national emissions reduction target aligned with international commitments such as the Paris Agreement. These targets are then broken down into sector-specific goals, considering the diverse contributions of industries such as transportation, energy, and agriculture to national emissions. For individuals, allowances are calculated based on average per capita emissions, currently estimated at 1.9 tons of CO_2 (or 2.8 tons of CO_2e) annually in India. Regional and demographic factors further refine these allowances, ensuring equity and practicality.

For corporations, quotas are derived based on industry-specific baselines. Heavy industries such as steel, cement, and chemicals are allocated higher allowances initially but are subjected to stricter reduction trajectories. Conversely, sectors like retail and service industries, which inherently produce lower emissions, receive proportionally smaller quotas. These allowances are adjusted for production volumes and operational efficiencies, encouraging innovation and sustainability.

For instance, a manufacturing company with annual emissions of 10,000 tons of CO_2e might be allocated an initial allowance of 8,000 tons, incentivizing a 20 percent reduction through the adoption of renewable energy and energy-efficient technologies. Similarly, a retail company operating primarily in urban settings may receive an allowance of 2,000 tons due to its relatively smaller carbon footprint, encouraging investments in energy-efficient logistics systems and renewable power sources. Such benchmarks not only encourage compliance but also drive technological progress across various industries.

Projected Impact on Emissions Reduction

The widespread adoption of the eCarbon Card system could lead to transformative reductions in carbon emissions. By incentivizing behavioral changes and fostering economic accountability, individuals and businesses are encouraged to adopt low-carbon practices. For example, a household switching to energy-efficient appliances, using public transport, and reducing waste could achieve a 10-15 percent reduction in its carbon footprint annually. On a national scale, such modest reductions per capita can translate into significant cumulative impacts. As of 2024, India's annual per capita emissions was 1.9 tons. India's recent ascent to the fourth-largest economy in the world has in part been driven by its rapid rise in urbanization. Its urban population in 2023 was about 510 million or ~ 35 percentage of the entire population. And it is the urban population compared to rural areas which is the major contributor to the annual per capita emissions. If one quarter of India's urban population, approximately 128 million people, each reduced their emissions by 0.75 ton annually, the result would be an annual reduction of 96 million tons of CO_2e — a monumental contribution toward India's climate goals.

Looking at China, another country who could be a good candidate for the adoption of eCarbon Card. In 2023, its urban population was about 911million or 64 percentage of the entire population — much higher than India's. If one quarter of China's urban population, approximately 228 million people, each reduced their emissions by 0.75 ton annually, the result would be an annual reduction of 171 million tons of CO_2e.

Industries, driven by both regulatory requirements and market incentives, can achieve even greater reductions. For instance, the steel sector, accounting for a significant portion of industrial emissions, could lower its footprint by transitioning to hydrogen-based production technologies. A case in point is Sweden's SSAB, which has already begun producing fossil-free steel[132]. India, with its large steel manufacturing base, can emulate such initiatives, potentially achieving a 15-20 percent reduction in emissions from this sector alone. Other high-emission sectors, such as power generation, can achieve similar results by scaling up renewable energy installations and implementing smart grid technologies to optimize energy distribution.

Thus, implementing a Carbon Credit system across individual, corporate, and community levels can reduce emissions by driving sustainable practices, funding green projects, and encouraging innovation. Here's a breakdown of how these reductions contribute to national targets by 2030:

1. **Individual Reductions**:
    - Potential Reduction: **140-200 million tons of CO_2** annually (assuming widespread adoption and behavior change).

2. **Industrial Reductions**:
    - Potential Reduction: **200-300 million tons of CO_2** annually through increased energy efficiency, renewable energy use, and carbon capture.

3. **Carbon Offset Projects**:
    - Potential Reduction: **100-200 million tons of CO_2** annually through forest conservation, regenerative agriculture, and waste-to-energy projects.

4. **Aggregate Impact**:
   - **Cumulative Reduction**: Approximately **500-700 million tons of CO_2** annually, which would account for a substantial portion of India's emissions reduction targets.

Economic Benefits for Households and Corporations

The economic advantages of the eCarbon Card system extend far beyond emissions reductions. For households, financial savings arise from reduced energy consumption, tax incentives, and rewards for sustainable practices. A middle-class family adopting solar panels, for instance, not only lowers its electricity bills but also earns additional income by selling surplus energy to the grid. Similarly, individuals who use the eCarbon Card to track and reduce their carbon footprints may receive rewards such as discounted utility rates, cashback on public transport, or tax credits for adopting sustainable technologies.

Corporations benefit from the opportunity to monetize emissions reductions through carbon credit trading. For instance, a logistics company optimizing its delivery routes and transitioning to electric vehicles could generate surplus credits, which can then be sold to companies exceeding their allowances. By providing such financial incentives, the eCarbon Card system encourages businesses to integrate sustainability into their core operations, leading to increased competitiveness and profitability in a green economy.

The ripple effects of these practices extend to the broader economy. By driving demand for green products and services, the system stimulates growth in sectors such as renewable energy, electric vehicles, and energy-efficient appliances. For example, the expansion of solar energy infrastructure in India has already led to the creation of thousands of jobs in manufacturing, installation, and maintenance. By incentivizing similar practices across industries, the eCarbon Card system has the potential to catalyze economic growth while simultaneously reducing emissions.

Table 7.1

Category	Example	Economic Impact
Households	Solar panel installation	20-30% reduction in electricity bills, income from selling surplus energy
Corporations	Logistics route optimization	Savings on fuel costs, revenue from surplus carbon credits
Renewable Energy Sector	Expansion of solar energy	Job creation in manufacturing, installation, and maintenance

Real-World Application: Personal Carbon Tracking and Rewards System

A pivotal component of the eCarbon Card system is the integration of a personalized carbon tracking and rewards mechanism. Through a dedicated app, individuals can monitor their carbon footprint and receive incentives for sustainable behaviors. This system fosters a culture of accountability and promotes a low-carbon lifestyle. Citizens can earn rewards such as discounts on eco-friendly products, lower public transport fares, or tax rebates for reducing their emissions.

Consider the case of Sweden's *Klimatkontot* (Climate Account), a tool enabling citizens to calculate their carbon footprint and receive personalized recommendations for reduction. Similar initiatives, such as Doconomy's DO Card, link carbon tracking to financial transactions, creating a seamless and actionable framework for individuals. In India, a similar program could encourage millions to adopt practices like carpooling, energy conservation, and the use of renewable energy. For example, an urban family reducing its reliance on private vehicles by switching to electric public transport could achieve a 15 percent reduction in transportation-related emissions.

Another potential application could involve using IoT-enabled smart meters that integrate with the eCarbon Card. For instance, an individual's electricity consumption could be directly linked to their carbon account, creating real-time feedback on usage patterns and providing immediate rewards for energy-efficient practices. Such dynamic systems encourage proactive engagement with carbon reduction goals.

Table 7.2

Behavioral Change	Incentive Mechanism	Impact
Reduced vehicle use	Discounts on public transport fares	10-15% reduction in urban emissions
Energy-effi-cient homes	Tax rebates, reduced utility bills	20-30% reduction in household emissions

Carbon Credit Mechanism: Driving Decarbonization

The carbon credit mechanism is central to the eCarbon Card system's success. By placing a tangible value on emissions reductions, it incentivizes both individuals and businesses to adopt sustainable practices. An individual staying within their annual carbon allowance can sell surplus credits, creating a financial reward for low-carbon living. Similarly, businesses investing in renewable energy or energy-efficient technologies can generate credits, offset their initial costs and driving further innovation.

An illustrative example is the European Union's Emissions Trading System (EU ETS), which has successfully reduced emissions across participating industries. By setting a cap on total emissions and allowing the trading of allowances, the EU has achieved significant reductions while maintaining economic growth. For instance, the ETS contributed to a 35% reduction in emissions from the power sector between 2005 and 2019. Applying similar principles in India, tailored to its unique socio-economic context, could yield comparable results. High-emission industries such as cement and steel can leverage the system to transition toward cleaner production methods.

Furthermore, small businesses can utilize the carbon credit marketplace to improve their sustainability. For example, a small textile manufacturer adopting energy-efficient machinery could generate credits, which could then be sold to larger corporations seeking offsets, thereby creating a new revenue stream for smaller players in the economy.

Table 7.3

Entity	Example Activity	Credits Earned
Individual	Staying below annual allowance	Sell surplus credits
Business	Investing in solar energy	Generate credits from savings
Heavy Industry	Adopting hydrogen technologies	Offset emissions through trading

Case Studies: Industrial Decarbonization and Community Engagement

The potential of the eCarbon Card system is further highlighted through case studies. In the industrial sector, companies like Tata Steel[133] have embarked on ambitious decarbonization journeys, integrating renewable energy and adopting innovative production methods. These initiatives demonstrate the feasibility of achieving substantial emissions reductions while maintaining competitiveness. For instance, Tata Steel's integration of waste gas recovery systems reduced emissions by 15 percent at key facilities, earning the company substantial carbon credits[134].

At the community level, projects such as afforestation and regenerative agriculture present additional opportunities. For example, the "Cauvery Calling" initiative in southern India promotes tree-based farming, sequestering carbon while enhancing agricultural productivity[135-137]. By integrating such projects into the eCarbon Card framework, communities can generate carbon credits, creating economic incentives for sustainable practices.

Similarly, a pilot program in rural Maharashtra demonstrated that transitioning from diesel to solar-powered irrigation reduced emissions by 40%, underscoring the potential for scalable impact[138].

A noteworthy international example comes from Kenya's Sustainable Energy Fund for Africa (SEFA), which funds small-scale renewable energy projects[139]. By leveraging carbon credits, local communities have been able to implement solar microgrids, providing affordable and clean energy to rural areas while earning carbon offsets that can be sold on international markets. Such models can serve as inspiration for Indian villages seeking to balance development with sustainability.

Table 7.4

Case Study	Initiative	Environmental Impact
Tata Steel	Renewable energy integration	20% reduction in operational emissions
Cauvery Calling	Tree-based farming	Sequestered 1 ton of CO_2 per hectare
Rural Maharashtra	Solar-powered irrigation	40% reduction in emissions
Kenya's SEFA	Solar microgrids	Carbon offsets for rural electrification

Mobilizing Capital Towards Clean Technologies

- **Demand for Low-Carbon Solutions**: As individuals and industries face carbon limits, the demand for low-carbon products and services rises. This stimulates growth in sectors like electric vehicles, renewable energy, energy-efficient appliances, and sustainable consumer goods.

- **Carbon Credit Revenue**: The trading of carbon credits can fund green projects by creating revenue streams for those investing in emissions reductions. Individuals, businesses, and communities investing in clean energy projects (like rooftop solar or electric vehicle charging stations) can earn carbon credits, further financing sustainable infrastructure.

- **Investment in Carbon Capture and Storage (CCS)**: Companies seeking to reduce their carbon footprint may invest in carbon capture, usage, and storage (CCUS) technology to generate additional credits, leading to large-scale reductions in industrial CO_2 emissions.

Impact on Emissions: As green projects expand, they provide sustainable energy sources and reduce the need for fossil fuels, leading to a direct reduction in CO_2e from the power and transportation sectors.

Driving Industrial Decarbonization

- **Corporate Carbon Credits**: Industries, which contribute significantly to India's emissions, would also have a cap on their emissions through carbon allowances. This incentivizes companies to adopt cleaner, more efficient technologies.

- **Shift to Renewable Energy**: Companies can meet their carbon credit targets by shifting to renewable energy sources like solar, wind, or bioenergy. By reducing reliance on fossil fuels, industries can generate carbon credits for excess reductions, which can be sold or reinvested.

- **Energy Efficiency Initiatives**: Companies that improve energy efficiency in manufacturing and operational processes can earn credits. For example, upgrading machinery to more efficient models or optimizing supply chains can result in measurable CO_2e reductions, creating additional revenue from credit trading.

Impact on Emissions: If India's industrial sector reduced emissions by even 15-20% due to carbon credit-driven initiatives, this would result in significant CO_2e reductions, given that industry contributes approximately 24 percent of India's total emissions.

Encouraging Carbon Offsetting Projects

- **Forest Conservation and Reforestation**: Carbon credits can support projects focused on afforestation, reforestation, and forest conservation, which are powerful carbon sinks. Local communities can participate in forest management or tree-planting programs, generating credits and preserving ecosystems.

- **Agricultural Carbon Sequestration**: Implementing regenerative agricultural practices, such as cover cropping, no-till farming, and agroforestry, allows farmers to sequester carbon in soils and generate credits. This improves soil health, increases productivity, and provides additional income through credit trading.

- **Waste-to-Energy and Biofuel Projects**: By converting organic waste into energy or biofuels, communities and businesses can reduce landfill emissions and generate carbon credits for sale, further supporting clean energy use.

Impact on Emissions: Carbon offset projects could account for hundreds of millions of tons in carbon reductions, as India has vast potential for land-based and agricultural carbon sequestration initiatives. Such projects also bring significant co-benefits, such as biodiversity preservation and rural job creation.

Catalyst for Green Market Development and Innovation

- **Increased Demand for Sustainable Products and Services**: As individuals and businesses seek to stay within their carbon allowances, demand for low-carbon products and services will rise. This includes electric vehicles (EVs), renewable energy, energy-efficient appliances, and green building materials.

- **Growth in Carbon Credit Market**: A new market for trading carbon credits would create a significant economic sector, with individuals and businesses trading credits based on surplus or deficit. This market can drive capital toward sustainable projects and carbon offset initiatives.

- **Incentive for Green Innovation**: Companies and startups will have incentives to innovate in areas like clean energy, energy efficiency, waste management, and carbon capture to help consumers and other businesses reduce their carbon footprint and minimize credit costs.

Economic Impact: India could see substantial growth in green sectors, with new business opportunities and job creation in renewable energy, electric vehicles, energy efficiency, and carbon management services. This growth could add billions to India's GDP by fostering a robust green economy.

Revenue Generation and Government Savings

- **Revenue from Carbon Credit Trades and Taxes**: The government can generate revenue through taxes on carbon credit transactions, as well as fees associated with the Carbon Card issuance and management. Additionally, auctioning carbon credits initially (before allocating) to industries could bring substantial funds.

- **Reduction in Health Expenditures**: Lower carbon emissions lead to better air quality, which can reduce healthcare costs related to respiratory and pollution-related illnesses. Improved public health can lead to savings in national healthcare spending.

- **Reduction in Fuel Import Costs**: By promoting lower fossil fuel consumption and incentivizing renewable energy use, the system could reduce India's dependency on imported oil and coal, improving the trade balance and reducing vulnerability to global fuel price fluctuations.

Economic Impact: The government could redirect savings from reduced healthcare costs and fuel imports toward renewable energy projects, infrastructure, and other growth areas. This redirection could also reduce India's fiscal deficit over time.

Job Creation and Workforce Transformation

- **New Jobs in Green Sectors**: The demand for low-carbon goods and services will increase employment in sectors like renewable energy, electric mobility, carbon trading, and environmental consulting.

- **Training and Reskilling Programs**: The transition to a low-carbon economy will necessitate reskilling and upskilling in green technologies. This could foster a highly skilled workforce, making India a leader in the global green economy.

- **Growth of Carbon Market and Financial Services**: As the carbon credit market matures, financial and regulatory services around carbon trading, credit management, and sustainable investing will grow, creating job opportunities in these sectors.

Economic Impact: With increased green jobs and workforce reskilling, India could see significant economic benefits, particularly in regions with strong renewable energy or natural resource potential. This shift could also lower unemployment rates in rural areas through agricultural carbon sequestration and reforestation projects.

Implementing the **Carbon Credit Concept** in India would require significant investment in infrastructure, technology, and administration, creating numerous job opportunities across various sectors. An estimated breakdown of potential job creation across different areas of the economy is provided in **Appendix XI – Job Creation Estimate**.

Overall Economic Impact

The implementation of the Carbon Credit concept could create approximately **1.2 million to 1.7 million jobs** across various sectors in India. Details of how this estimate is arrived at are provided in **Appendix XI**. This job growth would not only support the transition to a low-carbon economy but also stimulate economic development, particularly in rural and underserved regions. Additionally, these jobs would span a range of skill levels, from highly specialized roles in technology and finance to vocational roles in installation, agriculture, and community management.

By creating a diverse array of employment opportunities, the Carbon Credit system would contribute to sustainable economic growth while addressing climate change, positioning India as a leader in the global green economy.

Long-Term Vision: Aligning with National and Global Goals

The eCarbon Card system aligns seamlessly with India's climate commitments under the Paris Agreement and its goal of achieving net-zero emissions by 2070. By fostering accountability at both the individual and industrial levels, creating a market for carbon credits, and incentivizing innovation, the system provides a scalable model for balancing economic growth with environmental sustainability. Moreover, the integration of advanced technologies such as blockchain and AI ensures transparency and efficiency, addressing potential challenges in implementation.

In conclusion, the eCarbon Card system represents a paradigm shift in carbon management, combining environmental stewardship with economic opportunity. Through meticulous planning, robust infrastructure, and widespread engagement, this innovative approach has the potential to position India as a global leader in sustainable development. The examples and case studies presented herein underscore the system's viability and its capacity to drive meaningful change, making it a cornerstone of India's decarbonization strategy.

Chapter 8

Incentives and Rewards for Sustainable Choices,
and Benefits to the Government

The success of the eCarbon Card system hinges on its ability to effectively incentivize individuals, businesses, and communities to transition toward sustainable practices. By integrating financial benefits, social recognition, and community-driven initiatives, the system not only fosters a cultural shift but also creates tangible economic and environmental impacts. This chapter provides an expansive view of the mechanisms behind these incentives, enriched with real-world examples, potential implementation strategies, and the long-term societal benefits they could bring.

Financial Incentives and Benefits: Empowering Sustainability Through Economic Rewards

Financial incentives are one of the strongest motivators for driving behavioral change, as they provide immediate, tangible benefits to participants. Under the eCarbon Card system, financial rewards cater to individuals, businesses, and even communities, encouraging them to make low-carbon choices while reducing the economic burden of transitioning to sustainable lifestyles.

Rewards and Incentives could include -

- **Discounts and Vouchers for Green Purchases**: Reward users with discounts on eco-friendly products (for example, organic food, sustainable clothing, EV rentals) and carbon offsets for sustainable purchases.

- **Carbon Credits and Certificates**: Allow users who achieve substantial reductions to earn carbon credits that can be applied toward further rewards or offsetting their unavoidable emissions.

- **"Green Citizen" Recognition**: Users with consistent low-carbon scores could receive special recognition, like a "Green Citizen" status, with benefits such as lower property taxes or discounts on public transport.

For Individuals:

The eCarbon Card offers a variety of monetary incentives, such as cashback, discounts, and tax deductions for adopting green practices. A commuter, for example, who consistently opts for public transportation instead of driving a private vehicle could accumulate carbon credits, redeemable for discounts on transit fares or eco-friendly products. Consider the following example:

a resident of Mumbai reduces their daily commute emissions by switching from a car to the metro system, earning 100 credits monthly, which they redeem for discounts on energy-efficient appliances or a bicycle.

India has already seen success in incentivizing green energy at the household level. Under the *Mukhyamantri Solar Yojana* in Delhi, subsidies for rooftop solar installations have made renewable energy accessible to thousands of households, resulting in significant carbon reductions. This program could serve as a template for incorporating subsidies under the eCarbon Card system. For example, a household installing a 3kW solar system might save 30 percent on upfront costs through credits and see a 20–30 percent reduction in electricity bills annually.

Such schemes provide a blueprint for scaling financial incentives under the eCarbon Card system.

For Businesses:

Businesses, particularly those in high-emission sectors like logistics, manufacturing, and energy, stand to benefit immensely from the eCarbon Card system. Financial incentives for transitioning to renewable energy, electrifying vehicle fleets, or adopting energy-efficient production methods can lead to substantial cost savings via tax breaks and subsidies for adopting energy-efficient technologies, transitioning to renewable energy, or reducing emissions beyond regulatory requirements and additional revenue streams through tradable carbon credits.

A logistics company switching to electric delivery vehicles, for instance, can cut operational costs and generate surplus credits. Mahindra Logistics' transition to an electric fleet, coupled with route optimization, reduced CO_2 emissions by 1.2 KT (Kilo Tons) annually while cutting fuel expenses by 25 percent[140,141]. These savings could be further amplified under the eCarbon Card system by monetizing unused carbon allowances in a national carbon credit marketplace.

Table 8.1

Economic Impact of Low-Carbon Choices

Category	Example	Economic Impact
Households	Rooftop solar panel installation	20–30% reduction in electricity bills
Businesses	Transition to EV fleets	25% savings on fuel costs and tradable credits
Community Projects	Tree planting initiatives	Monetized carbon credits and improved local air quality

Carbon Credits and Offsetting: A Marketplace for Sustainability

At the core of the eCarbon Card system lies the concept of carbon credits, which monetize emissions reductions and foster a robust marketplace for sustainability. Individuals and businesses earn credits by engaging in low-carbon activities, such as energy efficiency upgrades, waste reduction, or sustainable transportation. These credits can be used to offset emissions or traded for financial rewards, creating an economic incentive to pursue greener choices.

Nationwide Reforestation Campaigns:

India's rural reforestation initiatives offer a compelling example of how offset programs can generate significant carbon sequestration benefits. Mangrove restoration projects in Kerala, for instance, not only absorb 30–50 tons of CO_2 per hectare annually but also improve coastal resilience — protection against natural disasters like cyclones and floods — and enhance livelihoods for local fishing communities[142]. Integrating such projects with the eCarbon Card could allow participants to directly support these efforts by purchasing carbon offsets tied to verified reforestation programs.

Consider the potential impact of a nationwide campaign: if 50 million citizens plant one tree each year and nurture it for five years, the resulting sequestration could offset approximately 1.23 million tons of CO_2 annually as detailed in **Appendix XII – Calculations**. Similarly, corporate solar farms that replace coal-powered plants could reduce up to 100,000 tons of CO_2 emissions per project, offering substantial credits to participating businesses.

Table 8.2

Examples of Carbon Offset Projects

Project	Impact	Annual CO_2 Sequestered
Rural Reforestation	Tree planting in rural areas	22 kilograms per tree
Mangrove Restoration	Coastal protection, livelihood support	30–50 tons per hectare
Corporate Solar Farms	Large-scale renew-able energy projects	100,000 tons per project

Social and Behavioral Incentives: Fostering a Cultural Shift

Social recognition and behavioral nudges play a critical role in fostering sustainable habits. The eCarbon Card system incorporates mechanisms like awards, gamification, and tiered reward systems to encourage individuals and organizations to adopt and sustain low-carbon lifestyles.

Digital Badges and Certifications:

Participants reducing their emissions significantly could earn digital badges or certifications, sharable on social media platforms or integrated into corporate branding. This public acknowledgment creates a sense of achievement and inspires others to follow suit. For instance, a housing society achieving a 40% reduction in collective emissions could be awarded a

"Green Community of the Year" title, enhancing its reputation and increasing property values.

Gamification and Tiered Rewards:

Gamification introduces an element of fun and competition to sustainability. Under a tiered reward system, participants progress through levels — bronze, silver, and gold — based on sustained emissions reductions. A family that reduces household emissions by 25% over a year could achieve the gold tier, earning exclusive discounts on eco-friendly vacations or access to green tech expos. This approach ensures continued engagement and fosters a culture of environmental stewardship.

International examples highlight the efficacy of such strategies. Japan's *Eco Points* program rewarded citizens with points redeemable for public transport tickets and energy-efficient appliances, achieving widespread adoption. Similarly, the UAE's *Green Rewards* program has successfully tied loyalty points to sustainable practices, creating an aspirational culture around environmental responsibility[143,144].

Table 8.3
Gamified Tier Rewards

Tier	Requirements	Rewards
Bronze	Basic sustainable practices	Discounts on public transport
Silver	Moderate emissions reductions	Cashback on eco-friendly products
Gold	Significant lifestyle changes	Invitations to sustainability expos and events

Community Incentives: Scaling Impact Through Collective Action

The eCarbon Card system amplifies its impact by incentivizing community-level initiatives, fostering collaboration for sustainability. Local neighborhoods participating in projects such as waste segregation, water conservation, or tree planting could earn collective credits, redeemable for funding green infrastructure or public amenities.

Success Stories:

Pune's *Zero Garbage Project* is a prime example of community-led sustainability[145-147]. Residents worked together to segregate and recycle waste, reducing landfill contributions and creating employment opportunities in waste management. Similarly, Bengaluru's *My Tree Campaign* incentivized urban afforestation by rewarding participants for planting and nurturing trees, enhancing the city's green cover[148,149].

Table 8.4
Community Initiatives and Benefits

Initiative	Impact	Rewards
Zero Garbage Project	Waste reduction, job creation	Community credits for green infrastructure
My Tree Campaign	Urban afforestation	Individual credits for sustainable actions

Government-Backed Subsidies and Tailored Incentives: Strengthening the Framework

Governments play a pivotal role in scaling the eCarbon Card system by providing subsidies for low-carbon alternatives, such as EVs, solar installations, and energy-efficient appliances. Tax rebates and grants for businesses investing in sustainable technologies further reinforce participation. Companies reducing emissions through renewable energy investments, waste reduction programs, or sustainable supply chains can earn green certifications, enhance their brand image and appeal to eco-conscious consumers. These certifications can be prominently displayed on products, websites, or marketing materials, reinforcing the company's commitment to sustainability.

For example, Tata Steel's initiative to integrate renewable energy and waste gas recovery systems into its operations not only reduced emissions by 15 percent but also earned the company significant recognition and financial benefits. Similarly, multinational corporations like IKEA have committed to becoming climate-positive by adopting renewable energy and circular economy practices, demonstrating the economic and reputational advantages of sustainability[150-152]. These examples underline the potential for businesses to gain a competitive edge by aligning with sustainability goals.

Challenges and Mitigation

Despite its promise, implementing the eCarbon Card system will require overcoming challenges such as funding reward programs, ensuring equitable access, and managing data privacy. Partnerships with eco-conscious brands, leveraging revenue from carbon credit trading, and integrating subsidies into the system can help address financial hurdles. Ensuring inclusiveness in reward structures and building robust data privacy measures are critical to the system's success.

Conclusion: Building a Low-Carbon Future

The eCarbon Card system demonstrates the transformative potential of integrating financial, social, and community incentives into a unified framework. By empowering individuals, businesses, and communities with rewards for sustainable behavior, it fosters a cultural shift toward decarbonization. With thoughtful implementation, robust infrastructure, and equitable design, the system can catalyze India's transition to a sustainable, low-carbon future.

Chapter 9

Viability, Challenges, Limitations, Potential Pitfalls, and Alternatives

An Honest Examination of the eCarbon Card System

The Vision Meets Reality

The eCarbon Card system is a bold attempt to reimagine climate action by weaving carbon accountability into the everyday lives of individuals and enterprises. Unlike top-down regulation or abstract market instruments, this system proposes a people-centered approach to decarbonization — one that recognizes individual agency, rewards sustainable behavior, and enables inclusive carbon trading. The system creates economic opportunities including substantial job creation. At a time when the world seeks practical, scalable, and just climate solutions, the eCarbon Card offers a distinctly innovative framework.

However, for such a system to move from concept to reality, it must prove viable not only on paper, but across diverse real-world conditions. Like any groundbreaking initiative, it comes with challenges and limitations that must be thoughtfully addressed. Its design must withstand political scrutiny, behavioral resistance, infrastructural constraints, and economic pressures. This chapter takes a deeper look at the system's feasibility and limitations, exploring what could go wrong, what needs to go right, and what we can learn from global and local examples. Potential alternatives are explored, and practical solutions are offered to strengthen the case for its implementation

Viability: Why the System Could Work

The eCarbon Card system's viability lies in its ability to align with India's climate goals, including its commitments under the Paris Agreement[5] and the ambitious target of achieving net-zero emissions by 2070. The system's success depends on several factors: public and corporate acceptance, robust technological infrastructure, regulatory support, and equality-driven implementation. By providing financial incentives and fostering behavioral change, the eCarbon Card system has the potential to transform society's approach to sustainability.

Public buy-in is critical to the system's success. Educating citizens about the importance of tracking and reducing their carbon footprints is essential. For example, Japan's "Eco Points" program[65] demonstrated how financial rewards could incentivize sustainable choices such as purchasing energy-efficient appliances. Similarly, the eCarbon Card system could reward individuals with credits for adopting public transportation, using renewable energy, or making other eco-friendly decisions. Over time, these incentives can create a societal shift towards sustainability.

Corporate participation is equally vital. Companies that

reduce emissions could earn carbon credits, which can be traded or used to offset their footprints. For instance, multinational corporations such as Tata Steel[89,133,134] and Infosys[153,154] have adopted strategies to integrate renewable energy and improve energy efficiency, aligning with India's broader climate objectives. The eCarbon Card system would provide additional incentives for such businesses, creating a ripple effect across industries.

Strong government support is indispensable. Policymakers must establish clear regulations, enforce carbon limits, and provide subsidies for green technologies. For example, the success of India's National Solar Mission, which facilitated the rapid expansion of solar energy capacity, illustrates how government-led initiatives can drive large-scale transformation. Integrating such programs with the eCarbon Card system would enhance its effectiveness by offering synergistic benefits.

Key Challenges and Mitigation Strategies

While the eCarbon Card system offers significant potential, its implementation involves several challenges that must be addressed to ensure success.

A. Privacy and Data Security

One of the most pressing concerns is data privacy. The system requires the collection of extensive personal and corporate data to track carbon footprints, raising concerns about data misuse. To address this, the system can leverage blockchain technology, ensuring secure and transparent data storage. For example, Estonia's e-Governance system has demonstrated how blockchain can protect sensitive information while maintaining transparency. Public awareness campaigns can further build trust by explaining how data will be used and safeguarded.

The sensitivity of corporate data also requires rigorous security protocols. Companies may hesitate to disclose emissions data due to competitive concerns. Here, encrypted storage and anonymized reporting can address confidentiality issues while maintaining accountability.

B. Economic and Social Equity

Equity remains a critical concern, as carbon quotas could disproportionately impact low-income individuals who may lack access to low-carbon alternatives. To mitigate this, the system can provide additional credits or subsidies to economically disadvantaged groups. For instance, low-income households could receive incentives to adopt cleaner cooking fuels or energy-efficient appliances. Programs like India's *Ujjwala Yojana*, which subsidized the adoption of LPG for rural households, highlight how targeted subsidies can drive equitable access to sustainable resources. Similar models can be adapted to provide allowances that prioritize essential needs without creating undue financial burdens.

C. Behavioral Resistance

Behavioral resistance to lifestyle changes presents a significant challenge. People may be reluctant to alter entrenched habits, such as driving private vehicles or consuming high-carbon goods. To address this, India's unique sociocultural context offers distinct opportunities for behavioral engagement. Public education campaigns, gamification, and financial rewards can encourage change — particularly when designed with culturally resonant motifs.

A tiered reward system — with bronze, silver, and gold levels — can appeal to India's deeply rooted cultural appreciation for status recognition, community-based honor, and milestone achievements. Similar structures are prevalent in Indian loyalty programs, public health campaigns, and even social media engagement. Such a system may not translate as effectively in Western contexts like the U.S., where individualism and privacy norms can diminish the influence of public-facing incentive schemes.

For example, households achieving significant emissions reductions could be rewarded with discounts on eco-friendly products or exclusive benefits such as subsidized access to green community events or public recognition through local governance platforms.

Behavioral economics also plays a role. Emphasizing immediate and tangible benefits — such as lower energy bills, improved indoor air quality, and health savings — can drive participation. These co-benefits may resonate more than abstract long-term climate goals.

While international examples like California's cap-and-trade system have shown that incentives can influence corporate behavior, their impact on individual-level behavior change has been more limited and less well-documented. In California, the cap-and-trade program generated billions in revenue for climate investments, but its direct influence on individual consumption behavior is debated. In contrast, India's strategy may require bottom-up engagement that leverages community identity, peer visibility, and social recognition — critical levers of behavior in many parts of the country.

D. Resistance from High-Carbon Industries

High-carbon industries such as steel, cement, oil and gas, and heavy manufacturing are likely to resist the implementation of the eCarbon Card system due to its potential impact on their operational costs, profitability, and market competitiveness. These industries often rely heavily on fossil fuels and have complex supply chains with significant carbon footprints, making it challenging for them to adapt quickly to stringent carbon allowances or financial penalties for excess emissions. Resistance may manifest in the form of lobbying against the policy, reluctance to adopt low-carbon technologies, or concerns about increased costs being passed on to consumers. Additionally, industries may argue that the system could disadvantage them in the global market, particularly if international competitors are not subject to similar carbon accountability measures.

To address this resistance, a strategic, phased approach is essential. One effective solution is to engage high-carbon industries early in the process, involving them in consultations and pilot programs to understand their concerns and build a sense of ownership in the system's implementation. Establishing a collaborative dialogue with industry stakeholders can help identify sector-specific challenges and co-develop tailored solutions, such as gradual reduction targets and transitional support. For instance, industries could be provided with longer timelines for compliance, allowing them to plan and invest in sustainable technologies without disrupting operations.

Another key strategy is to offer targeted incentives to support the transition to low-carbon alternatives. These incentives could include tax breaks, subsidies for adopting green technologies, or access to low-interest loans for energy efficiency upgrades. For example, cement manufacturers could receive financial support for integrating carbon capture and storage (CCS) technologies into their production processes, while steelmakers could be incentivized to adopt hydrogen-based production methods. By reducing the financial burden of transitioning to sustainable practices, the eCarbon Card system can encourage industries to view decarbonization as an opportunity rather than a threat.

Additionally, creating sector-specific carbon benchmarks and peer comparison frameworks can drive accountability and competitiveness. High-performing companies that meet or exceed reduction targets could be publicly recognized, enhancing their reputation and market appeal. This could also encourage lagging companies to improve their performance to remain competitive. Moreover, establishing a carbon credit marketplace linked to the eCarbon Card system would allow industries to offset some of their emissions by purchasing credits from lower-emission sectors or entities. This flexibility provides a transitional pathway while incentivizing overall reductions in emissions.

E. Corporate Manipulation and Greenwashing

Corporate greenwashing, or exaggerating sustainability efforts, is a significant risk. Strict monitoring, third-party audits, and penalties for non-compliance can mitigate this. Transparent reporting mechanisms, such as public emissions disclosures, can enhance accountability. The EU Emissions Trading System's requirement for verified emissions reports serves as an effective model for ensuring corporate compliance. Additionally, blockchain technology can help track corporate carbon transactions, ensuring authenticity and reducing the likelihood of manipulation.

F. Technological and Administrative Costs

Developing and maintaining the technological infrastructure for the eCarbon Card system involves significant costs. Public-private partnerships can address these challenges. For example, India's Unified Payments Interface (UPI) system was developed through collaboration between the government and private entities, demonstrating the feasibility of cost-sharing models. Pilot programs in regions with higher digital literacy can also reduce initial costs while providing valuable insights for scaling the system.

The administrative burden of managing allowances and credits across millions of users is another challenge. Advanced automation, AI-powered analytics, and machine learning algorithms can streamline these processes. For instance, AI models can identify fraudulent activities (e.g. using classification ML algorithms), optimize credit allocations, and provide personalized recommendations for reducing emissions.

Limitations and Pitfalls: What the System Cannot (Yet) Do

Even with the best design and intentions, there are inherent limitations.

First, accurate *carbon accounting* is still an evolving science and very complex requiring accurate measurement of emissions across diverse activities. Life-cycle assessments vary by geography, product, and usage. Consumption-based emissions are hard to quantify at the household level without intrusive monitoring. This can be addressed incrementally—starting with high-confidence data sources (for example, electricity bills, fuel receipts) and refining overtime through machine learning and user feedback. Leveraging machine learning algorithms can improve data accuracy by analyzing patterns in energy use, transportation, and consumption. For example, predictive analytics used in smart grids have improved energy efficiency by optimizing power distribution based on consumption trends.

Second, *price volatility* in the carbon marketplace could create instability, lead to black markets or discourage participation. If credit prices swing too widely, trust erodes. Introducing regulatory measures such as price floors, ceilings, and government-backed reserve credits can prevent sharp imbalances, using circuit breakers like stock markets, launching reserve pools of credits during spikes and offering forward contracts to businesses to hedge against risk. Additionally, setting aside reserve credits for emergencies can prevent shortages during periods of high demand. China's national emissions market uses such controls, balancing environmental ambition with market discipline. Its carbon market implemented government-issued credits to balance supply and demand providing a useful example of how price stability can be achieved. Multi-year or multi-month rolling averages, such as used by California's ETS, could address price volatility.

Third, the *digital divide* cannot be ignored. Millions in rural India remain offline or digitally marginalized. For the system to be inclusive, it must work via SMS, USSD, and offline kiosks — not just smartphones. Literacy barriers must be overcome with iconography, audio instructions, and local-language support.

Fourth, *political cycles* may threaten continuity. The system must be anchored in legislation, insulated from populist swings, and governed by an independent authority like a proposed Carbon Authority of India (CAI). This entity could include representation from civil society, science, business, and government, ensuring pluralistic oversight.

Global Comparisons and Adaptable Alternatives

The viability of the eCarbon Card system is supported by real-world examples of successful carbon management initiatives. South Korea's Emissions Trading Scheme (ETS), for instance, incentivizes corporations to reduce emissions through sector-specific caps and trading mechanisms. The program has successfully reduced emissions in heavy industries while fostering innovation in low-carbon technologies. The program's success highlights the potential of market-based approaches to drive sustainability.

In India, renewable energy programs such as the National Solar Mission have driven significant emissions reductions. By integrating such programs with the eCarbon Card system, participants could earn additional credits for adopting solar energy solutions, further enhancing the system's impact. Programs like National Solar Mission or the LPG-focused *Ujjwala* scheme demonstrate how behavioral, and infrastructure shifts can be driven through public engagement and incentives. The eCarbon Card builds on these lessons but adds personalization and tradability — something previous policies lacked.

EU ETS is one of the world's most mature and comprehensive carbon pricing mechanisms. It sets a cap on total emissions from power plants, industrial facilities, and airlines operating within the European Economic Area, and allows these entities to trade emission allowances in a regulated market. Its phased rollout and stringent verification protocols have established credibility, transparency, and measurable reductions in GHG emissions. Importantly, the EU ETS demonstrates how digital registries, robust monitoring, and cross-border trading can form the backbone of an efficient carbon market. These lessons in compliance enforcement and transparency can be adapted for the eCarbon Card system at the individual level, ensuring trust and accountability in a decentralized carbon market.

Ghana's Cookstove Credit Program, part of its broader clean cooking initiative, offers a compelling model for integrating rural, low-emission behaviors into national and international carbon markets. The program certifies improved biomass and clean cookstoves distributed in rural areas and quantifies their carbon savings to generate tradable carbon credits. These credits are then sold in voluntary or compliance markets, bringing tangible financial flows into underserved communities. Ghana's approach illustrates the feasibility of using verifiable low-tech interventions for generating carbon credits and highlights the importance of community involvement, government backing, and simple monitoring tools — all of which are critical for scaling the eCarbon Card system in rural India.

Table 9.1 compares the impact of and learnings from various programs.

Table 9.1

Example Program	Impact	Lessons Learned
South Korea ETS	Reduced emissions in heavy industries	Sectoral caps and flexible trading mechanisms
National Solar Mission (India)	Rapid expansion of solar energy	Government subsidies and strong targets
EU Emissions Trading	Cross-border trading of carbon allowances	Verified reporting mechanisms
Ghana's cookstove credits[155]	Rural projects as legitimate credit generation tools	Government-backed reward programs build loyalty

Alternatives to the eCarbon Card system, such as carbon taxes and direct subsidies, offer complementary tools. Taxes offer simplicity but risk being regressive. Subsidies work but can be expensive. Though these approaches lack the personalized accountability of the eCarbon Card, combining them with the system can create a comprehensive framework for carbon reduction. For instance, a hybrid model that includes both carbon credits and direct financial incentives could address diverse needs while promoting sustainable behaviors. The system is not a silver bullet, but when paired with targeted policies, it becomes a powerful engine for systemic change.

Cultural and Psychological Considerations

For the successful adoption of the eCarbon Card system, cultural and psychological factors are just as important as technological infrastructure. At the heart of this lies the principle that trust — not technology — determines adoption. People are more likely to embrace new systems when they feel respected, empowered, and included, rather than surveilled or penalized. Therefore, the eCarbon Card must be framed not as a tool of restriction or carbon austerity, but as a symbol of "carbon dignity" — an instrument that honors sustainable lifestyles and responsible choices. Positive cultural cues can reinforce this framing: awarding green badges, celebrating "solar credit heroes," or issuing community recognitions for climate leadership can all elevate the card as a marker of progress, not punishment. A powerful example of this comes from Brazil's Bolsa Verde program, which compensated rural families for preserving forest land. It succeeded not by emphasizing carbon metrics, but by telling a story of stewardship, dignity, and pride[156]. The eCarbon Card must follow a similar path — rooted in values, identity, and agency — if it is to inspire behavioral change across India's diverse social and economic fabric.

Projected Economic and Environmental Impact: Numbers That Matter

The economic and environmental impact could be immense. By incentivizing green investments, the system can create new revenue streams and job opportunities in sectors such as renewable energy, energy efficiency, and sustainable agriculture. For example, as per IRENA, the renewable energy sector in India has already generated over 720,000 jobs — employing approximately 1.02 million people in 2023, reflecting significant growth in green jobs across the country[157-159]. The expansion of the eCarbon Card system could further boost employment in green industries. Implementation of the Carbon Card System

On the environmental front, widespread adoption of the eCarbon Card system could lead to significant emissions reductions. If just 50 million Indians reduce their emissions by 10 percent annually, the country could save 100 million tons of CO_2e — equivalent to taking 20 million cars off the road. A well-run carbon marketplace could create over a million new green jobs in logistics, clean energy, agriculture, and analytics. Solar and wind adoption could rise by 30 to 40 percent, especially if tied to credit multipliers. Similarly, businesses adopting sustainable practices could achieve collective reductions that align with national and global climate goals. Table 9.2 summarized the potential key impact.

Table 9.2

Metric	Potential Impact
Annual Citizen Reductions	100 million tons of CO_2
Job Creation	Over 1 million new green jobs
Renewable Energy Adoption	40% increase in solar and wind installations

These are not just estimates. They are achievable benchmarks — if the system is designed for adaptability, inclusivity, and trust.

Conclusion

The eCarbon Card system represents a bold and innovative approach to address climate change. By integrating financial incentives, technological innovation, and public participation, it offers a comprehensive solution for achieving carbon reduction goals. However, the eCarbon Card is not without flaws. But its strength lies in its ability to evolve, adapt, and integrate. It does not claim to replace all climate policies. Instead, it offers something few systems have managed to do translate the abstract notion of a carbon footprint into a tangible, accountable, and actionable reality. While challenges such as data privacy, equity, and technological complexity must be addressed, the system's potential benefits far outweigh its limitations.

If implemented thoughtfully, with transparency, empathy, and scientific rigor, the eCarbon Card could become more than a policy tool. It could become a symbol of shared responsibility — where every citizen, every company, and every action counts and can serve as a model for global climate action. Its ability to democratize carbon management and foster a culture of sustainability positions it as a transformative tool in the fight against climate change. As nations worldwide seek solutions to the decarbonization challenge, the eCarbon Card system offers a viable and scalable pathway to a sustainable future.

As India pilots this vision, the world will watch. And if successful, the eCarbon Card could become a blueprint not just for climate governance — but for a new kind of climate citizenship.

Chapter 10

*Next Steps and the Future of Decarbonization
with the eCarbon Card System*

The eCarbon Card system represents a transformative leap in addressing the dual challenges of climate change and sustainability. As a tool for democratizing carbon management, this system offers a promising path forward by integrating technology, policy, and behavioral shifts. However, its potential impact relies on the structured implementation of next steps and a clear vision for the future of decarbonization. This chapter explores the roadmap for advancing the eCarbon Card system, identifies the roles of key stakeholders, and envisions a global shift toward sustainability that transcends national borders.

Scaling the eCarbon Card System

Scaling the eCarbon Card system requires a phased approach that balances experimentation with large-scale deployment. The initial phase of implementation should focus on pilot programs to refine the system's mechanics and demonstrate feasibility. These pilots, conducted in select urban and rural regions, can serve as laboratories for testing the system's core components, such as carbon tracking, credit allocation, and trading mechanisms. For example, a pilot program in a metropolitan city like Bengaluru, coupled with a rural counterpart in Karnataka, could yield insights into the system's adaptability to diverse socioeconomic contexts. Metrics such as public participation rates, emission reductions, and the efficiency of the credit marketplace would help fine-tune the system for broader rollout.

Furthermore, these pilot programs can involve partnerships with academic institutions, local non-profits, and government bodies to evaluate both quantitative and qualitative impacts. For instance, organizations like The Energy and Resources Institute (TERI) in India can contribute to evaluating emission reductions. Given its credibility, multidisciplinary expertise, and long-standing influence on policy and industry, TERI could play a critical role in the design, validation, and implementation of the eCarbon Card System. TERI has shaped numerous national strategies (for example, India's Energy Policy, the Perform Achieve Trade (PAT) scheme), and has long-standing policy engagement credentials. For example, TERI could develop carbon footprint models for individuals, MSMEs, and sectors tailored to Indian lifestyles and economic activity; validate and update India-specific emission factors for electricity, transport, food, housing, and other categories and lead the methodology for calculating carbon allowances based on region, income, and access to infrastructure. TERI has already conducted deep carbon assessments across industries and states (e.g., the "India Greenhouse Gas Program")[160] and has developed regionally differentiated emission profiles. There could be a role for NITI (National Institution for Transforming India) *Aayog* as well.

In addition to pilots, partnerships with local governments

and organizations will play a critical role. For instance, state governments could collaborate with non-profits and educational institutions to train local administrators and promote public awareness. Successful examples, such as the decentralized implementation of India's *Swachh Bharat Abhiyan*, demonstrate the power of localized approaches in scaling national initiatives.

Subsequent phases should involve gradual scaling, starting with state-wide adoption before expanding nationally. This phased approach allows for iterative improvements while mitigating potential risks. Drawing lessons from South Korea's Emissions Trading Scheme (ETS), which began with a few high-emission sectors before expanding, India can ensure that the eCarbon Card system is robust and effective.

Table 10.1

Phases of Implementation	Key Activities	Examples
Pilot Programs (Year 1*)	Testing in urban and rural settings	Bengaluru and rural Karnataka
State-Level Expansion (Year 2)	Regional adoption and training programs	Collaboration with state governments
National Rollout (Year 3)	Full-scale deployment with monitoring	Lessons from South Korea's ETS

*the starting year of the phased implementation of the *eCarbon Card* system

Long-term Projections and Potential Impact

One of the primary long-term goals of the eCarbon Card system is to implement progressive reductions in carbon allowances for individuals and businesses. This incremental approach is critical for driving continuous emissions reductions. By gradually lowering the permissible carbon footprint over time and increasing the cost of excess credits, the system incentivizes behavioral shifts and technological adoption, fostering a deeper commitment to sustainability. For example, the allowances for urban residents with access to public transportation and renewable energy infrastructure may be reduced more aggressively than those in rural areas, balancing equity with environmental goals. Such reductions not only push for individual and corporate accountability but also ensure India stays on track to meet its climate targets, including net-zero emissions by 2070.

Assuming gradual uptake, here's how this system could scale:

1. **Initial Years:**
    »   With an estimated adoption by 10 percent of India's population, around 140 million users could cut their emissions by 1 ton each annually through conscious choices, resulting in a 140-million-ton reduction in national emissions.

2. **Scaling Impact:**
    Given India's population size, even a small reduction per capita can translate to significant emission reductions.

» If 10 percent of India's population (currently at 140 million people) reduces their carbon footprint by just 1 ton per year through sustainable habits, it could result in a total reduction of 140 million tons of CO2, which is a meaningful contribution to national goals.

» If adoption expands to 30 percent of the population over time, with a target reduction of 1-1.5 tons per user per year, this initiative could potentially lower emissions by 400-600 million tons annually.

3. **Behavioral Shift**:
   » Beyond the quantified reductions, the program can lead to long-term behavioral change, making sustainable choices a social norm. This shift can also lead to systemic impacts as citizens advocate for cleaner infrastructure, more efficient public transit, and stricter environmental policies.

Leveraging Technology for Seamless Implementation

Technology forms the backbone of the eCarbon Card system. A sophisticated digital platform is essential for tracking emissions, managing carbon credits, and facilitating transactions. Blockchain technology offers a secure and transparent solution for maintaining records, while cloud computing ensures scalability. Estonia's e-Governance system), which integrates blockchain to manage citizen data securely, serves as a model for the eCarbon Card's digital infrastructure. This system has not only improved transparency but also fostered public trust, an essential element for the success of the eCarbon Card.

Artificial intelligence (AI) and machine learning (ML) can enhance the system's efficiency by analyzing user behavior and predicting carbon footprints. These technologies will be vital to monitor, predict and manage the reduction of carbon emissions. They can personalize recommendations for reducing emissions, such as suggesting energy-efficient appliances or public transportation options. For instance, AI-driven recommendation systems like those used by e-commerce platforms can be adapted to nudge users toward sustainable choices. Additionally, the integration of Internet of Things (IoT) devices, such as smart meters, can provide real-time data on energy consumption and emissions, enabling accurate tracking and quicker adjustments.

The integration of geospatial technologies can add another layer of utility to the system. For example, geographic information systems (GIS) can be employed to analyze regional emission patterns, enabling policymakers to make data-driven decisions about infrastructure investments and resource allocation. A practical use case could be mapping high-emission zones in urban areas and targeting them with specific interventions like better public transport options or renewable energy installations.

Technology also plays a crucial role in creating a user-friendly experience. A mobile application could serve as the primary interface, allowing users to monitor their carbon footprints, trade credits, and access sustainability tips. Cloud-based data storage ensures seamless access across devices while safeguarding privacy. The development of APIs to integrate with financial services and utility providers can streamline data collection, making the system more efficient.

Strengthening Public Engagement and Awareness

The success of the eCarbon Card system hinges on public buy-in. Effective communication strategies are crucial for educating citizens about the system's benefits and encouraging participation. Public awareness campaigns, such as those used to promote India's *Swachh Bharat Abhiyan*, can inspire collective action. These campaigns should emphasize the tangible benefits of sustainable choices, such as cost savings and improved air quality, to resonate with diverse audiences.

Public awareness efforts must also incorporate digital and social media platforms to reach younger generations. Short, engaging videos explaining the benefits of the eCarbon Card and testimonials from early adopters could create a ripple effect. For example, the use of digital platforms to promote solar rooftop installations in Gujarat significantly boosted adoption rates, demonstrating the power of targeted online campaigns[161].

Gamification and reward systems can further enhance engagement. For instance, users who stay within their carbon quotas could earn points redeemable for discounts on sustainable products or services. Cities like Amsterdam have successfully used similar strategies to promote cycling by offering incentives to participants in bike-to-work programs[162]. Such approaches can foster a culture of sustainability while making the transition enjoyable and rewarding.

Addressing Equity and Inclusion

Equity is a cornerstone of the eCarbon Card system. Ensuring that the system benefits all citizens, regardless of income or location, is critical for its success. Special provisions should be made for economically disadvantaged groups and rural communities with limited access to low-carbon alternatives. For example, subsidies for renewable energy adoption or cleaner cooking fuels can help bridge the gap.

The system must also account for regional variations in carbon footprints. Urban areas with extensive public transportation networks may require lower allowances than rural areas reliant on traditional fuels. Tailoring the system to these differences ensures fairness and inclusiveness, fostering public trust and widespread adoption.

An innovative approach to equity could involve community-level carbon pooling. For instance, rural communities could collectively earn credits through sustainable practices such as afforestation or water conservation, which could then be redistributed among members.

Encouraging Corporate Participation

Corporations play a pivotal role in the success of the eCarbon Card system. By aligning corporate strategies with the system's objectives, businesses can drive large-scale emission reductions while benefiting from new economic opportunities. For instance, companies that exceed their emission reduction targets can sell surplus credits in the marketplace, creating a financial incentive for sustainability.

Case studies from multinational corporations illustrate the potential impact. For example, Unilever's Sustainable Living

Plan has demonstrated how integrating sustainability into business operations can reduce emissions while enhancing profitability[163,164]. Similarly, Tata Steel's commitment to achieving net-zero carbon emissions aligns with the goals of the eCarbon Card system, highlighting the synergy between corporate efforts and policy initiatives.

Corporate participation can also be enhanced through partnerships with startups specializing in green technologies. For example, collaborations with companies developing energy-efficient materials or carbon capture solutions could accelerate progress toward decarbonization goals.

Long-Term Goals and Adaptability

Its adaptability and progressive nature are integral to ensuring continued relevance and effectiveness in a dynamic global climate landscape. The system's success depends on a phased approach that reduces carbon allowances over time, integrates with global carbon markets, and embraces technological and policy advancements to ensure resilience and scalability.

The system's adaptability also includes aligning with global carbon markets and standards. Over time, the eCarbon Card could connect to international emissions trading platforms, enabling India to participate in cross-border carbon credit trading. Such integration would provide economic benefits by allowing Indian citizens and businesses to sell surplus credits internationally, fostering a global culture of carbon accountability. For instance, lessons could be drawn from the European Union's Emission Trading System (EU ETS), which has successfully incentivized emissions reductions across member states. Aligning the eCarbon Card system with global standards not only boosts its credibility but also positions India as a significant player in international climate negotiations, strengthening its leadership on climate action.

Another essential aspect of the system's longevity is continuous monitoring and feedback loops. Regularly reviewing emissions data and user feedback allows for timely refinements in allowances, incentives, and penalties. This ensures that the system remains responsive to evolving climate goals and societal needs. Advanced technologies like artificial intelligence and machine learning could analyze patterns in carbon usage, providing insights into areas where interventions are most needed. For example, if data reveals high emissions from specific sectors or regions, the government could introduce targeted policies, such as renewable energy subsidies or stricter regulations, to address these issues. Such an adaptive framework ensures that the system remains effective in achieving its overarching goals.

Futureproofing

To ensure resilience in the face of future challenges, the eCarbon Card system must be future-proofed. This involves preparing for advancements in technology, shifts in energy landscapes, and potential changes in policy or public behavior. For instance, as renewable energy becomes more accessible and affordable, the system can adjust its allowances and incentives to reflect reduced carbon intensity in energy consumption. Similarly, the integration of emerging technologies like blockchain and IoT would enhance data accuracy, transparency, and security, ensuring the system remains robust against misuse or obsolescence. By proactively planning for such changes, the eCarbon Card system can maintain its relevance and effectiveness over decades.

Global Implications and Future Prospects

The eCarbon Card system has the potential to serve as a model for global decarbonization efforts. As climate change transcends national boundaries, international collaboration is essential. India's leadership in implementing the eCarbon Card system could inspire other nations to adopt similar approaches, fostering a global culture of sustainability.

International partnerships, such as the Global Carbon Project[165], can facilitate knowledge sharing and resource pooling. By collaborating with countries that have implemented carbon trading systems, such as the European Union or China, India can leverage best practices and avoid potential pitfalls.

Moreover, the system's adaptability to diverse contexts makes it a valuable tool for addressing global challenges. For instance, small island nations vulnerable to rising sea levels could adopt localized versions of the eCarbon Card system to manage their carbon footprints and build resilience.

Envisioning a Sustainable Future

The long-term vision for the eCarbon Card system extends beyond emissions reduction. By fostering a culture of sustainability, the system can drive innovation, create economic opportunities, and improve quality of life. Imagine a future where low-carbon lifestyles are not only a necessity but also a source of pride and fulfillment. Communities could thrive on renewable energy, green jobs, and sustainable practices, paving the way for a resilient and equitable world.

This vision is not merely aspirational; it is achievable with collective effort and commitment. The eCarbon Card system represents a step toward realizing this future, offering a practical and scalable solution to one of humanity's greatest challenges. Table 10.2 is reproduced to highlight the potential impact for India with the adoption of the card.

Table 10.2

Metric	Potential Impact
Annual Citizen Reductions	400 million tons of CO_2e (2028), 500 million tons of CO_2e (2030)
Job Creation	Over 1 million new green jobs
Renewable Energy Adoption	40% increase in solar and wind installations

Finally, the successful implementation of the eCarbon Card system would position **India as a global leader in carbon accountability**. By demonstrating that a nation with vast socio-economic diversity can adopt and scale such an ambitious initiative, India can inspire other countries to implement similar systems. This leadership would not only enhance India's global reputation but also encourage international collaboration on climate solutions. For example, India could share its experiences and best practices with other developing nations, fostering a collective effort to combat climate change. Additionally, India's leadership could attract investments in green technologies and infrastructure, further accelerating its transition to a low-carbon economy.

Conclusion

The eCarbon Card system is more than a policy initiative; it is a call to action for individuals, corporations, and governments to unite in the fight against climate change. Its potential to transform society's approach to sustainability is immense, but realizing this potential requires thoughtful implementation, robust technology, and widespread engagement. By incorporating progressive reductions, aligning with global markets, embracing adaptability, and demonstrating leadership, the system paves the way for a sustainable future

As India takes bold steps toward decarbonization, the eCarbon Card system can serve as a beacon of hope and innovation for global climate action. By addressing challenges, fostering collaboration, and embracing a shared vision for the future, we can build a sustainable world for generations to come.

eCarbon Card
India's Smart Climate & Economic Solution

CORE DIFFERENTIATORS

REWARDS-BASED PERSONAL CARBON TRACKING & TRADING

* Annual allowance in carbon units
* Deducts based on lifestyle activities
  - transport, energy use, consumption
* Unused credits can be sold or redeemed
* Low emitters earn, high emitters buy offsets, incentivized

MARKET-FIRST
Built on economic incentives, not moral imperatives

RURAL-FIRST
Rewards India's low-carbon lifestyles

WHY INDIA? WHY NOW?

* Climate impacts are real and urgent
* India has unique digital public infrastructure
* 700 M citizens live low-carbon lifestyles

DIGITALLY INCLUSIVE
Accessible via app, kiosk, or SMS

GREEN ECONOMIC ENGINE
Creates jobs in agriculture, retrofits, and clean energy

IMPACT POTENTIAL

* Carbon market at the base of the pyramid
* Transparency
* De-centralized
* Decarbonization
* Behavioral shift

SCALABLE
Easy integration with Aadhaar, UPI, state schemes

Final Pitch

eCarbon Card – India's Smart Climate & Economic Solution

The eCarbon Card is a technology-driven, reward-based carbon tracking and trading system for every Indian citizen and business.

It creates personal carbon allowances, deducted based on real-world behavior — transport, energy use, food, and consumption — and enables trading of unused credits.

This is not a punitive scheme. It's a performance-based, gamified, and market-aligned tool that:

✓ Rewards individuals for sustainable living
✓ Supports rural incomes and digital inclusion
✓ Drives green innovation and behavior change
✓ Avoids ideological framing — making it apolitical, scalable, and investor-ready

Why Now?

✓ Climate urgency is real — affecting crops, water security, urban heat, and economic stability.
✓ India has a unique opportunity to lead with a pragmatic, incentive-driven decarbonization model.

✓ Digital public infrastructure (Aadhaar, UPI, ONDC) enables rapid, inclusive rollout.

✓ New green jobs and rural income are essential for political and social stability in a decarbonizing economy.

What is the eCarbon Card

A personal or enterprise-linked carbon wallet issued annually to individuals and businesses. Key features:

✓ Annual allowance in carbon units (e.g., an average of 1.5 tons/year per person) — like a mobile data plan

✓ Deductions based on lifestyle or business activities (e.g., car use, electricity, fertilizer)

✓ Surplus credits can be sold, traded, or redeemed

✓ Low emitters earn; high emitters buy offsets

✓ Works offline (SMS/kiosk) and online (app/portal)

It's climate-smart economics — for individuals and business entities.

It's like a Paytm for carbon. But instead of spending rupees, you spend carbon credits — and if you save them, you earn.

Core Differentiators

✓ Market-first: Built on economic incentives, not moral imperatives

✓ Rural-first: Rewards India's low-carbon traditional lifestyles

✓ Digitally inclusive: Works via app, kiosk, or SMS

✓ Green economic engine: Catalyzes demand for EVs, solar, compost, biogas, retrofits

✓ Scalable: Easy integration with Aadhaar, UPI, state schemes

Why This Works for India

Unlike in the West, where sustainability is getting politicized and tied up with DEI narratives, India can chart a different path — rooted in:

✓ Self-reliance

✓ Rural empowerment

✓ Digital infrastructure

✓ And climate-smart livelihoods

India's farmers, artisans, and low-income families already live within nature's limits.

Why not pay them for it?

Indian urban consumers want to make better choices. Why not guide them with simple feedback and incentives?

What This Unlocks (Impact Potential)

✓ Domestic Carbon Market at the Base of the Pyramid: Driven not by global offset schemes but local action; 700M+ rural Indians become net credit sellers. A green micro-economy powered by citizens.

✓ Urban emissions transparency: Citizens and businesses internalize the cost of emissions

✓ Decentralized decarbonization: Villages earn credits

for climate-smart agriculture, composting, and
reforestation

✓ A surge in green jobs: Youth agents, auditors,
data verifiers, clean tech trainers, solar installers,
agroforestry workers

✓ A rise in green entrepreneurship: Retrofits, biogas,
sustainable agriculture

✓ Behavioral shift: Gamified app nudges consumers
toward sustainable decisions

✓ And above all — a mass movement where
decarbonization is not a burden, but a badge of honor.

What is needed are policy champions, investors, and phil-
anthropic partners to:

✓ Launch multi–district national pilot

✓ Co-develop the tech stack (app, APIs, carbon tracking
backend, trading platform)

✓ Partner with state governments, cooperatives, and
climate-tech innovators

✓ Support public awareness and carbon
literacy campaigns

✓ Test pricing, carbon categories, and trading rules
in real-time

India doesn't need to copy Western models. It can define its
own path — using tech, incentives, and inclusion to decarbonize
without polarization.

India may be the first country in the world where decarbonization is as easy and rewarding as using a digital wallet. The eCarbon Card can be India's greatest green innovation — Where the poor earn, the rich adapt, and the country wins.

Epilogue

A Future Within Reach

As we turn the final pages of this book, we find ourselves not at an ending, but at a beginning.

The eCarbon Card is more than a policy or a technological solution - it is an invitation to re-imagine the relationship between people, planet, and prosperity. In India, with its vast diversity, deep-rooted frugality, and cutting edge digital infrastructure, this opportunity strongly resonates a climate challenged century - trust, transparency, and collective action - measurable and tractable.

Let us build a world where carbon is not just a cost to avoid, but a value to create. Let us democratize decarbonization.

Let us make every citizen count - and every gram of carbon matter.

The future is not somewhere we are going. It is something we are co-creating - today, together.

Let the journey begin.

"We did not come to fear the future - we came here to shape it"

Epilogue

A Future Within Reach

"We do not inherit the earth from our ancestors;
we borrow it from our children."
— Native American Proverb

As we turn the final pages of this book, we find ourselves not at an ending, but at the beginning.

The eCarbon Card is more than a policy proposal or a technological solution — it is an invitation to reimagine the relationship between people, planet, and prosperity. It is a blueprint for a world where every citizen becomes not just a consumer of resources, but a conscious steward of carbon — earning, savings, and spending it wisely, just as we do with money, time, and energy.

In India, with its vast diversity, deep-rooted traditions of frugality, and cutting-edge digital infrastructure, this idea is not just plausible — it is powerful. From the smallholder farmer in Bihar to the tech entrepreneur in Bengaluru, from a tribal community in Odisha to a retiree in Pune, the eCarbon Card has the potential to unite us in a common purpose: building a climate-smart society that is just, inclusive, and thriving.

But turning this vision into reality requires more than policy or code. It requires trust. It requires partnership. It requires the courage to try something new, and the humility to adapt it along the way.

This is not a silver bullet. It will not replace the need for structural reforms, clean energy transitions, or international cooperation. But it can become the glue that binds those solutions together at the individual and community level. It can drive awareness, accountability, and spiration — from the bottom up.

As we move deeper into a climate-challenged century, the most valuable currencies will not be oil or gas, but trust, transparency, and collective action. The eCarbon Card is one such currency — measurable, tradable, and meaningful.

Let us build a world where carbon is not just a cost to avoid, but a value to create.

Let us democratize decarbonization.

Let us make every citizen count — and every gram of carbon matter.

The future is not somewhere we are going.

It is something we are co-creating — today, together.

Let the journey begin.

Critical Questions, Candid Answers

A Conversation Around the eCarbon Card Concept

Over the course of discussions, talks, and pilot engagements around the eCarbon Card, several key questions have consistently come up along with many though-provoking comments. This section compiles and addresses some of those most frequently asked — to clarify, deepen, and extend the conversation and thus providing further insight into the idea and its potential impact.

Q1. Given 700M people in India are poor or live in rural areas, how does the idea of eCarbon Card help them on a day-to-day basis?

A. Getting this right is essential to making the eCarbon Card not only effective for decarbonization but also inclusive, equitable, and empowering, especially for India's 700 million+ low-income citizens living in rural areas.

Let's break it down into a detailed and holistic response, focusing on why, how, and what the eCarbon Card can do for this demographic.

Why the eCarbon Card Matters for the Rural and Poor Population

India's rural and low-income citizens are:

- The most vulnerable to climate change (heatwaves, floods, crop failure).

- The least responsible for historical emissions.

- Often excluded from digital or financial innovations.

The eCarbon Card, when designed with empathy and accessibility, can become a climate justice tool — not just a carbon tracking mechanism — with the aim to:

- Empower rural households economically

- Reward sustainable practices they already follow

- Protect them from market distortions

- Promote low-carbon development

How the eCarbon Card Helps: Everyday Benefits for Rural and Poor Citizens

1. Direct Carbon Allowance = New Economic Asset
   » The government allocates yearly carbon allowances to every citizen (say, 1–2 tons per person, for example, 1.2-1.5 tons per person for urban and 1.8-2.0 tons per person for rural population).

» Since rural and poor citizens emit far less than wealthier urban dwellers, they will have surplus carbon credits.

» These credits can be sold or traded on a government-backed carbon marketplace.

» Result: A *new income stream* simply for living sustainably — rewarding traditional low-carbon lifestyles (cycling, walking, solar cooking, rain-fed farming).

Example: A family in rural Bihar uses a biomass stove, walks to school, and doesn't own a car. Their emissions are well below the limit. They earn ₹600/month from selling unused carbon credits to an urban resident who exceeds their allowance.

2. Incentives for Traditional & Sustainable Practices
    » Many rural practices are inherently sustainable — organic farming, reusing goods, community transport, and low-energy lifestyles.

    » The eCarbon Card system can:
      • Recognize these practices as low-carbon.
      • Reward them with a bonus, carbon credits or financial incentives.
      • For farmers, this could be tied to sustainable agriculture certifications or green farming bonuses.

Example: A millet farmer in Rajasthan switches to drip irrigation. He receives extra carbon credits as a reward — either in cash or in-kind (free solar pump, seeds, and so on).

3. Cashless Micro-Transactions Through Carbon Credits
    » eCarbon Cards can act like a prepaid card or wallet, usable in carbon-approved local stores, markets, or government services.

    » Carbon rewards can be redeemed for:
        • Discounts on fertilizers, cooking fuel, seeds
        • Free or subsidized solar lanterns, biogas kits, school meals, or public transport
        • This creates localized economic circulation driven by sustainable behavior.

Example: A family earns 100 carbon credits for using a biogas stove. They use it to get a ₹500 discount on school uniforms or kerosene at a government ration shop.

4. Access to Green Subsidies and Schemes via Carbon Score
    » The eCarbon system could link to welfare schemes:
        • PM KUSUM (solar pumps)
        • PM Ujjwala (LPG)
        • PDS (Public Distribution System) subsidies
        • MGNREGA (Mahatma Gandhi Rational Rural Employment Guarantee Act)

    » Those with low carbon footprints or verified sustainable behaviors can be prioritized for benefits, promoting climate-smart development.

Example: A tribal village with low per capita carbon usage gets prioritized for a solar microgrid pilot.

5. Climate Resilience and Energy Security
    » By encouraging solar, clean cooking, energy-efficient appliances, the eCarbon Card improves energy access and resilience to climate shocks.

    » Community carbon savings could fund:
        • Local solar pumps
        • Cold storage for crops
        • Flood barriers or tree planting

Example: A village collectively saves 100 tons of CO_2 via green practices. The community uses these credits to get funding for a solar-powered water purification unit.

6. Jobs and Digital Inclusion in Rural Areas
    » The rollout of eCarbon Cards creates millions of local jobs:
        • Data entry agents at eCarbon kiosks
        • Carbon field verifiers (like MNREGA supervisors)
        • Sustainability coordinators (to train households in earning carbon rewards)

    » It will also expand digital literacy and access to technology, especially for women and youth.

Example: A village in Odisha trains 10 young women to help households report emissions, track rewards, and trade credits — creating income and leadership opportunities.

What's Needed to Make This Work for Rural and Poor Citizens

To make the eCarbon Card system truly inclusive and beneficial for India's rural populations and economically marginalized citizens, it must be designed with the realities of access, infrastructure, and agency in mind:

1. Simplicity and Accessibility
   - » The Card should function offline and via SMS or USSD, not just smartphones.
   - » Local eCarbon kiosks in villages and mandis should enable manual data entry and support.
   - » Participation should not depend on literacy or bank access; biometric or Aadhaar-linked options can simplify use and verification.

2. Trust and Transparency
   - » Clear info on how carbon is calculated and how people benefit.
   - » Third-party NGOs or panchayats involved in verifying low-carbon behaviors.
   - » Regular community audits, grievance redressal systems.

3. Localized Carbon Accounting
    » Emissions should reflect regional contexts:
        • Biomass stove ≠ same as LPG
        • Bullock cart ≠ same as scooter
    » Local emissions benchmarks help ensure fairness.

4. Built-In Protections
    » No penalizing excessive carbon use in low-income households (for example, during emergencies).

    » Cap credit prices to prevent market manipulation.

    » Provide free credits for essential needs like healthcare, childbirth, education-related travel.

Conclusion: A Tool for Justice, Not Just Emission Tracking

The eCarbon Card, if designed with equity in mind, has the power to:

• Reward the poor for what they already do well — live sustainably.

• Empower rural communities with new income and recognition.

• Bridge the rural-urban divide by letting those with excess emissions pay those with low ones.

- Develop grassroots green employment opportunities and infrastructure

- Build climate resilience, where it's needed the most.

In essence, the eCarbon Card can be more than a carbon tracker — it can be a platform for rural economic revival, social equity, and climate-smart development for the 700 million who must be central to India's climate journey.

Q2: Can you clarify further by providing an example of how a model village in India can benefit by adopting this?

A: Here's a visual story + model village example to bring the concept of the eCarbon Card for rural India to life — told through the lens of a fictional yet realistic village. This narrative can serve as a prototype or demonstration project for policymakers, innovators, or communities exploring climate-equity solutions.

Model Village: *Suryagram* — A Net-Zero Village in Bihar
Location: Rural Bihar
Population: 1,200
Occupation: Mostly small farmers, weavers, masons, and local service workers
Literacy: Moderate, digital access via community center
Poverty level: High, but strong community ties and traditions of sharing

Scene 1: The Launch of eCarbon Card

The District Collector visits *Suryagram* to launch a pilot of India's eCarbon Card program.

- Every adult in the village receives an eCarbon Card linked to their Aadhaar number.

- A **local kiosk** is set up in the panchayat bhavan, run by trained youth to help villagers:
  - » Log farming inputs, cooking fuels, and transport habits

  - » Check their carbon balances

  - » Redeem carbon rewards

Ramvati Devi, a 55-year-old widow, is helped to log her use of a chulha (biomass stove) and her seasonal farming practices. She doesn't own a vehicle or use electricity for irrigation.

Scene 2: Carbon Footprint Assessment

Each villager's carbon footprint is estimated using:

- **Type of cooking fuel** (biomass vs. LPG)
- **Transport use** (walking, bullock carts, public buses)
- **Electricity usage** (solar panels vs. grid)
- **Farming practices** (organic inputs, rain-fed fields)

Ramvati's annual footprint is just 0.4 tons of CO_2e — well below the average national allowance of 1.5 tons.

Scene 3: Rewards for Sustainable Living
Because Ramvati is under her carbon limit:

- She earns 1.1 tons worth of surplus carbon credits

- The credits are automatically uploaded to her eCarbon Card

- She has two choices:
    1. Sell them (through the kiosk) to an urban user who has exceeded their limit
    2. Redeem them for in-kind benefits

She chooses to redeem 0.5 credits for a solar lamp for her grandson's studies and saves the rest to sell when prices go up.

Scene 4: Community Market and Incentives
Villagers who stay under their carbon limits receive:

- Discounts on kerosene, fertilizers, or school uniforms

- Priority for green subsidies (e.g., solar pumps)

- Invitations to training in organic farming, which earns bonus credits

Ramvati joins a group of women switching to vermicompost, earning 0.2 credits/month and learning how to grow vegetables for sale.

Scene 5: Transport and Youth Employment

- The eCarbon program (Carbon Authority of India) hires 10 local youth as:
  - » Carbon field data collectors
  - » Digital support agents
  - » Awareness volunteers

- *Suryagram* gets an e-rickshaw partially funded by the community's shared carbon savings, used to transport kids to school and elders to the health clinic.

Scene 6: Suryagram Goes Net-Zero

After 18 months:

- *Suryagram*'s total emissions = 200 tons CO2e
- Carbon credits earned = 270 tons CO2e
- The village becomes carbon negative and sells surplus credits to a nearby city.
- The revenue is pooled to fund:
  - » A solar water pump
  - » Biogas for the Anganwadi (childcare center)
  - » A tree-planting campaign for flood resilience

Scene 7: Replication and Expansion

News spreads. Other villages request eCarbon Cards.

- The district scales it to 100 villages.

- Women's self-help groups are trained to co-manage village carbon data.

- Schools begin to teach students how to earn and track carbon credits through everyday actions like planting trees, recycling, or using cycles.

Table I
Impact Summary (1 Year in *Suryagram*):

Metric	Value
Average Carbon Footprint/Person	0.6 tons/year
Credits Earned by Village	320 tons
Carbon Income Generated	₹9.6 lakhs
Households Earning Regular Credit Income	85%
Youth Employed via eCarbon Ecosystem	10
Renewable Energy Installed	2 solar pumps, 1 microgrid
Women Self-Help Groups Active	4
Government Subsidies Unlocked	₹5 lakhs
Net Environmental Impact	Net-zero

What's the Vision?
If 10,000 villages like *Suryagram* adopt this model:

- Millions of tons of CO_2e can be avoided or offset
- Crores in new income can be generated for the rural poor

- India can build a bottom-up decarbonization movement rooted in equity and community resilience

Q3: How would the Urban vs. Rural Impact of the eCarbon Card compare in India?

A: The answer is presented in the form of a Briefing Paper:

Briefing Paper: Urban vs. Rural Impact of the eCarbon Card in India

Background

India, with a population of 1.4 billion+, faces a dual challenge:

- Rapid urbanization and rising consumption driving up emissions.
- 700 million+ rural and poor citizens contributing the least but bearing the brunt of climate impacts.

The eCarbon Card proposes a personal and business-level carbon allowance system, enabling individuals to track, trade, and reduce their carbon emissions, while incentivizing sustainable behavior.

Objective of Comparison

To assess how the implementation of eCarbon Cards would affect urban vs. rural populations, considering:

- Emission patterns
- Economic opportunities
- Behavior change
- Equity implications

Table II
Urban vs. Rural Impact Scenarios

Dimension	Urban Scenario	Rural Scenario
Average Emissions per Capita	2.5–4.0 tons CO_2e/year	0.5–1.2 tons CO_2e/year
Typical Sources of Emissions	Private vehicles, ACs, appliances, meat, flights	Cooking fuels, irrigation, limited transport, basic lighting
Likely Status Under eCarbon Card	Net carbon buyers (over emitters)	Net carbon sellers (under emitters)
Behavior Change Pressure	High (need to shift to EVs, public transport, green homes)	Low or none (already low-carbon lifestyle)
New Economic Opportunity	Must buy extra credits = added cost	Can earn income by selling surplus credits
Implementation Challenge	Digital tracking, resistance to change	Infrastructure, education, access
Equity Impact	Encourages accountability among high emitters	Rewards and empowers low emitters
Offset Generation Potential	Limited (dense space, high consumption)	High (tree planting, composting, low-tech solutions)

Rural Empowerment through eCarbon Card

- **Income generation**: Poor or low-emitting households can earn credits and sell them in carbon markets.
- **Access to green tech**: Credits redeemable for solar lanterns, biogas, clean stoves, or efficient pumps.
- **Recognition of sustainable practices**: Traditional organic farming, cycling, reusing, composting are now *rewarded behaviors*.

Example: A rural family using a biomass stove and rain-fed farming may emit just 0.6 tons/year. With a 1.5 ton/year allowance, they can earn and sell up to 0.9 tons of CO_2e credits annually — creating a new micro-income stream.

Urban Transformation Incentivized

- **Carbon usage transparency**: High-emission lifestyles (frequent flyers, SUV users, large homes) become visible and accountable.
- **Green choices rewarded**: Public transport, plant-based diets, renewable electricity reduce deductions and conserve credits.
- **Peer pressure + market incentives**: High earners are nudged to adopt sustainability for financial and social reputation benefits.

Example: An urban couple with two cars, central AC, and air travel exceeds their limit. They must either pay for offsets or shift behaviors — switch to metro, use solar, eat less meat.

Economic Impact

- **Urban**: Higher probability of exceeding carbon allowance, leading to purchase of additional credits. Encourages sustainable behavior and green lifestyle shifts.
- **Rural**: Most rural households will remain below their carbon caps. They can monetize surplus carbon credits, generating a new income stream. This system can uplift rural livelihoods without punitive pressure.

Behavioral Incentives and Lifestyle Shifts

- **Urban Citizens**:
  - » Incentivized to adopt energy-efficient appliances, use public transport, and shift to renewable energy.
  - » High-frequency users of air travel, ride-hailing services, and high-consumption lifestyles face higher credit usage.
- **Rural Citizens**:
  - » Rewarded for maintaining low-carbon traditions (walking, cycling, rain-fed farming).
  - » Transition support to cleaner cooking (e.g., biogas) is reinforced by carbon credits.

Policy & Design Recommendations

To ensure the eCarbon Card delivers *differentiated yet fair outcomes*:

1. **Progressive Allowances:** Base carbon allowance on lifestyle and needs, not just flat per capita.
2. **Rural Carbon Aggregation:** Let communities pool unused credits for village-level income or projects (e.g., solar microgrids, water pumps).
3. **Smart Differentiation:** Cooking fuel use in poor households ≠ AC use in upper-income homes — differentiate carbon cost per context.
4. **Digital + Analog Access:** Cards must work via SMS, voice, and village kiosks — not just apps or smart devices.
5. **Green Credits for Essential Needs:** Offer "zero-carbon" exemptions for travel to school, hospital visits, or public services in poor/rural areas.

Implementation Challenges and Solutions

- Digital Divide: Offline tools (SMS-based entry, local kiosks) can ensure rural access.
- Verification: Community-based audits and local governance institutions (Panchayats) can help validate emissions data.
- Market Fairness: Regulatory oversight is needed to prevent manipulation of carbon credit prices.

Conclusion

The **eCarbon Card**, if implemented thoughtfully, could:

- Empower rural India with new income and dignity
- Nudge high-emitters in cities to embrace low-carbon lifestyles
- Drive behavior change and demand for sustainable products
- Build a just, decentralized decarbonization pathway for India
- Shift the climate conversation from penalty to participation

Q4: What are the biggest sources of carbon emissions in India and how does implementation of eCarbon Card attack them differentially?

A: Understanding India's biggest carbon emission sources is key to evaluating whether the eCarbon Card concept can make a meaningful impact, and whether it does so equitably or uniformly.

Let's break it down into two parts:

Part 1: India's Major Sources of Carbon Emissions

India's total greenhouse gas emissions (as of 2024) are approximately 4.1 billion tons (GT) of CO_2e/year, making it the 3rd largest emitter globally — though per capita emissions (India's population in 2024 ~ 1.451 billion) are still low (2.8 tons of CO_2e per person/year vs. 15+ tons in the U.S. and global average of 7 tons).

Table III
India's Top Emission Sources (by sector):

Sector	Share of National CO2e Emissions	Key Emission Drivers
1. Energy production	~40%	Coal-fired power plants, grid electricity
2. Industry (cement, steel, etc.)	~20%	High-heat industrial processes
3. Transport (road, rail, air)	~10%	Diesel/petrol vehicles, trucks, flights
4. Agriculture	~14%	Methane from rice, livestock, fertilizer use
5. Buildings	~6%	Urban HVAC, lighting, appliances
6. Waste	~4%	Landfills, sewage, open burning
7. Residential cooking and heating	~3%	Biomass, LPG, coal stoves
8. Other (LUCF*, forestry)	~3%	Deforestation and land degradation

*Land Use, Land Use Change

Observation: Emissions are highly concentrated in industrial, energy, and transport sectors — but millions of small emitters across households also contribute cumulatively.

Part 2: How the eCarbon Card Aligns with These Sources

Does the eCarbon Card attack emissions *proportionally* or *differentially*?

Answer: Both — but with intentional *differentiation* to preserve fairness, while ensuring high emitters pay more.

Let's walk through the sectors again and examine how eCarbon Card affects them:

1. Energy Production (40%)

Structural/Upstream Emissions

eCarbon Card Effect: Indirect + Market Pressure

- Electricity consumers (households, offices, businesses) are assigned a carbon value based on their energy source.
- Coal-heavy users are penalized with higher deductions.
- Encourages people to switch to solar power, demand cleaner grids.

Differentiated approach: Doesn't penalize poor grid users in villages but incentivizes rooftop solar or efficient appliances in urban homes.

2. Heavy Industry (20%)

Corporate/Institutional Emissions

eCarbon Card Effect: Limited for individuals, Direct for businesses

- For MSMEs (Micro, Small, and Medium Enterprises) and businesses, eCarbon Cards can track usage of cement, steel, and chemicals.

- Those with high embedded emissions consume more credits and must offset or pay more.
- Drives demand for green building materials, low-carbon cement/steel.

Differentiated approach: Allows local manufacturers to earn credits by switching to green production.

3. Transport (10%)

Private and commercial vehicle use

eCarbon Card Effect: Strong and Direct

- Each vehicle trip is logged with GPS/fuel data or inferred from payment data (e.g., fuel purchases, Uber rides).
- Personal transport becomes a major part of an individual's carbon balance.
- High car/motorbike usage = faster depletion of carbon credits.

Differentiated:

- A city dweller with multiple cars will need to buy extra credits.
- A rural cyclist or bus user earns carbon income.

4. Agriculture (14%)

Fertilizers, rice, livestock, irrigation

eCarbon Card Effect: Farmer-centric + Reward-based

- Farmers input data on fertilizer usage, irrigation type, and crop choice.
- Low-emission practices (organic, solar pumps, rainfed crops) are rewarded with extra credits.

- Over time, it encourages a shift to climate-smart agriculture.

Differentiated: Smallholder farmers with sustainable practices are net credit sellers, while chemical-heavy industrial farms may become net buyers.

5. Buildings (6%)

Electricity, AC, appliances, elevators

eCarbon Card Effect: Per-person tracking in urban settings

- Tracks individual electricity use and housing size/type.
- Urban dwellers in energy-guzzling buildings use more credits.
- Incentivizes adoption of LEDs, solar water heaters, insulation.

Differentiated:

- A high-rise penthouse pays more than a rural mud home.

6. Waste (4%)

Landfills, open burning, poor segregation

eCarbon Card Effect: Community-based

- Credits can be earned for waste segregation, composting, recycling, especially at the community or panchayat level.
- Urban housing societies can form carbon co-ops for clean waste practices.

Differentiated

- Promotes low-waste habits among rich, while rewarding traditional frugality in poorer communities.

7. Residential Cooking and Heating (3%)

Stoves, fuelwood, LPG

eCarbon Card Effect: Immediate, Tangible Impact

- Carbon usage varies based on stove type: LPG > biomass > biogas > solar cooker
- Households can log their cooking fuel monthly and receive bonus credits for low-emission systems.

Differentiated:

- A tribal woman using a solar cooker can earn credits, while a city gas user spends theirs faster.

8. Land Use & Deforestation (3%)

Deforestation, urban sprawl, degraded soils

eCarbon Card Effect: Offset Opportunities

- Individuals or communities can plant trees or restore land to generate carbon offsets.
- Encourages village-level carbon farming or eco-tourism zones.

Differentiated:

- Rural areas can sell offsets to urban buyers to balance national footprint.

Summary Table IV
Emission Source vs. eCarbon Card Impact (India)

Dimension	Urban Scenario	Rural Scenario
Average Emissions per Capita	2.5–4.0 tons CO_2e/year	0.5–1.2 tons CO_2e/year
Typical Sources of Emissions	Private vehicles, ACs, appliances, meat, flights	Cooking fuels, irrigation, limited transport, basic lighting
Likely Status Under eCarbon Card	Net carbon buyers (over emitters)	Net carbon sellers (under emitters)
Behavior Change Pressure	High (need to shift to EVs, public transport, green homes)	Low or none (already low-carbon lifestyle)
New Economic Opportunity	Must buy extra credits = added cost	Can earn income by selling surplus credits
Implementation Challenge	Digital tracking, resistance to change	Infrastructure, education, access
Equity Impact	Encourages accountability among high emitters	Rewards and empowers low emitters
Offset Generation Potential	Limited (dense space, high consumption)	High (tree planting, composting, low-tech solutions)

Key Takeaway: The eCarbon Card uses a *differentiated impact model* — focusing strongly on sectors where individual and community behavior change is possible, while leaving structural emitters (like power plants) to broader policy instruments.

Final Insight: Differentiation Is a Feature — Not a Flaw

The eCarbon Card is inherently differentiated — by design — to:

- Protect the poor
- Reward the sustainable
- Encourage behavior change in high-emission groups

It doesn't attack all emitters proportionally (uniformly) because that would be unfair and could hurt the very groups that are already climate-resilient and low-emitting.

Instead, it's a precision instrument — attacking the problem where it's biggest, while uplifting those who live lightly on the Earth.

Q5: How would one account for gradual but robust yearly increase in India's population while determining per capita allowances - progressively decreasing with time to drive decarbonization?

A: When designing a per capita carbon allowance system like the *eCarbon Card*, India's growing population and declining carbon budget both need to be accounted for simultaneously and equitably. Here's how this can be done:

The key is to implement a dual-indexed allowance system that:

1. Adjusts downward over time to meet India's national carbon targets (e.g., Net Zero by 2070)
2. Adapts to a growing population without unfairly squeezing individuals

1. Set a National Carbon Budget Trajectory

- Determine India's total allowable emissions year by year (for example, 2.60 Gt CO_2 in 2026 → 2.40 Gt CO_2 in 2030 → 1.7 Gt CO_2 in 2040, etc.)
- This budget should decline over time in line with international commitments (for example, India's NDCs[63,64], Paris Agreement[5])

2. Forecast Population Growth

- Use projections from the UN or India's Census Bureau
- India's population is expected to grow from ~1.43 billion (2024) to ~1.55 billion (2040+)

3. Calculate Annual Per Capita Carbon Allowances

Per Capita Allowance =
National Carbon Budget ÷ Population

As the population rises and the budget drops, per-person allowances naturally decline.

Example:

Table V

Projected Decline in Per Capita Carbon Allowance vs. Total Carbon Budget India

Year	Total Carbon Budget (Gt CO_2)	Population (Billions)	Per Capita Allowance (tons CO_2)
2022	2.69	1.425	1.89
2024	2.75	1.451	1.90
2025	2.90	1.460	1.99
Phased Introduction of eCarbon Card System begins in 2026			
2026	2.60	1.500	1.73
2030	2.40	1.525	1.57
2035	2.05	1.579	1.30
2040	1.70	1.623	1.05
2050	1.00	1.680	0.60
2070	0.00	1.750	0.00

Here is the visual dashboard showing year-by-year projections for India's:

- **Per Capita Carbon Allowance** (in tons CO2 per person)
- **Total National Carbon Budget** (in Gt CO2)

Fig. I

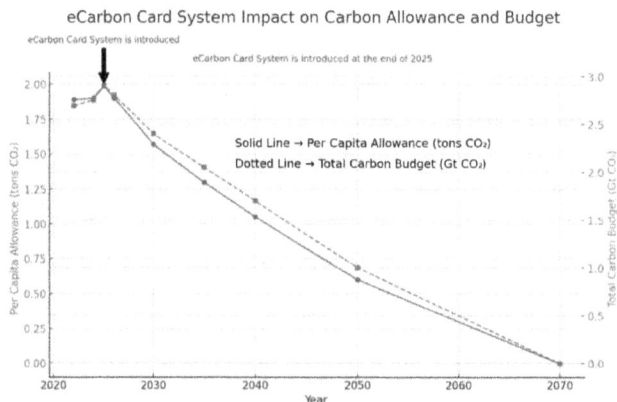

eCarbon Card System Impact on Carbon Allowance and Budget

As the chart illustrates:

- The carbon budget steadily decreases toward 2050 to align with climate goals.
- At the same time, the growing population reduces per-person allowances, driving the need for individual and systemic decarbonization

Q6: How can the eCarbon Card work in a country as diverse and complex as India?

A: Diversity is a feature, not a flaw. The eCarbon Card is built on tiered allowances, context-sensitive carbon accounting, and localized implementation models. For instance, a tribal farmer in Odisha will have a very different carbon profile and usage pattern from an urban commuter in Delhi — and the system is designed to reflect that. It leverages India's existing digital infrastructure (Aadhaar, UPI, DigiLocker) to enable scale and flexibility.

Q7: Won't this penalize the poor who have fewer choices?

A: Quite the opposite. Poor and rural citizens typically have much lower carbon footprints and will often become net credit earners. For them, the eCarbon Card is an economic empowerment tool, offering income from surplus carbon credits, access to clean technology, and recognition of their inherently sustainable lifestyles.

Q8: Who determines how much carbon each action emits? Isn't that subjective?

A: Carbon accounting uses internationally recognized Life Cycle Assessment (LCA) methods and verified emissions factors. For India, these factors will be customized regionally, and continually improved through open data, third-party validation, and machine learning models based on real-world use.

Q9. Is There a Precedent or Accepted Range?

Yes. Several countries and researchers have experimented with personal or sectoral carbon budgeting. While no universal number exists, models often reference:

- Global per capita carbon budget to stay within 1.5°C (~ 2–3 tons CO_2e/year)

- India's current average (~1.9 tons/person/year) as a baseline

- International sectoral benchmarks from the IEA, IPCC, and Science-Based Targets Initiative (SBTi)

The eCarbon Card adapts these global standards to local realities, using a tiered and just transition approach.

Q10: How is the price of carbon credits determined in this system?

A: Prices will be market-driven, influenced by demand, supply, and sectoral benchmarks. There can be tiered pricing structures and government floor prices to prevent volatility. Individuals and communities with surplus credits can choose to sell at market rate or to institutional buyers, like corporates or governments seeking offsets.

Q11: What happens if someone runs out of credits mid-year?

A: They can buy additional credits from others in the marketplace, or from the government at a slightly higher rate (to encourage peer-to-peer trading first). Essential needs like medical care or school commutes are exempt or subsidized. The system is supportive, not restrictive.

Q12: Can't people game the system or fake low-carbon behaviors?

A: Like any economic system, there is risk. But with smart auditing, blockchain-backed transactions, community-level verification, and randomized checks, fraud can be minimized. Plus, trusted institutions (panchayats, cooperatives, SHGs) will help validate data in rural areas.

Q13: How does the eCarbon Card apply to businesses or corporations?

A: Business entities can receive sector-specific carbon budgets, and track emissions through procurement, production, and logistics data. SMEs that stay under budget can earn tradable credits or receive preferential treatment in green public procurement, while larger emitters are encouraged to offset via community projects or innovation.

Q14: How does this compare to a carbon tax or cap-and-trade system?

A: The eCarbon Card is more personal, decentralized, and behavioral. Unlike carbon taxes, which are top-down and financial in nature, the eCarbon Card fosters bottom-up participation, localized rewards, and social recognition — making decarbonization a lived experience rather than an abstract policy.

Q15: Isn't this too ambitious or idealistic?

A: Big changes often start as ambitious ideas. The eCarbon Card doesn't need to work perfectly everywhere on Day 1. It can begin with targeted pilots, voluntary adoption, and gradual scale-up — just like UPI, Aadhaar, and MGNREGA did. The vision is big, but the execution is incremental, inclusive, and iterative.

Q16: How do we ensure trust in the system?

A: Trust comes from transparency, local participation, and tangible benefits. The eCarbon platform should be open-source, auditable, and governed by a multi-stakeholder council that includes citizens, scientists, policymakers, and technologists. Just like financial credit scores, carbon profiles must be protected and owned by the individual.

Q17: Can this really help India meet its climate targets?

A: Yes — by creating massive behavioral shifts, reducing demand for high-emission goods, and scaling community-level offsets. Even a 10 percent shift in behavior among 500 million people can result in hundreds of millions of tons of CO_2e saved annually. The eCarbon Card is not a silver bullet, but it's a critical piece of the climate puzzle.

Q18: How does the eCarbon Card support a green economy?

A: By creating new income streams, carbon jobs, and market demand for sustainable goods and services. Entrepreneurs can build apps, tracking tools, and green products; youth can become carbon auditors or advisors; local governments can monetize community credits for development.

Q19: What role can citizens play in making this a reality?

A: Everything. This system thrives on public participation, feedback, and trust. Citizens can — sign up for pilot programs, organize community climate clubs, help family and neighbors understand their carbon choices and demand accountability from companies and governments

This is not a system *for* people. It's a system *by* and *with* people.

Q20. In today's global context where climate and equity policies are increasingly politicized, how can the eCarbon Card gain support across different political, social, and economic landscapes?

A. As climate policies, equity frameworks, and sustainability narratives encounter growing political sensitivities in different parts of the world, it is important to position solutions like the eCarbon Card in ways that transcend ideology and resonate with a broad spectrum of stakeholders — from policymakers and businesses to everyday citizens.

The eCarbon Card's strength lies in its versatility of framing. It is not anchored to any single worldview. Instead, it can be introduced through multiple narratives — each appealing to a different constituency — while staying true to the core goal of democratizing carbon accountability. Here's how:

1. Reframe the Narrative: Focus on Empowerment

Instead of presenting the eCarbon Card as a tool of regulation or restraint, position it as:

- A personal empowerment tool that rewards choices and lifestyles
- A market-based innovation that allows individuals and communities to earn from sustainable behavior
- A platform where carbon frugality becomes financial opportunity, especially in low-emission households

2. Market Innovation, Not Mandate

Governments and businesses are more likely to adopt systems that:

- Encourage voluntary participation
- Function like a carbon wallet or digital budgeting tool
- Reward rather than punish — "emit less, earn more"

This makes it a productivity tool for citizens, not a compliance burden.

3. Anchor in Local and National Aspirations

The eCarbon Card aligns with national development priorities:

- Energy independence through incentivized clean energy adoption
- Local job creation in climate tech, infrastructure, data, and agriculture
- Digital India and grassroots empowerment by leveraging existing digital platforms (Aadhaar, UPI, ONDC)
- Reflects cultural values like thrift, simplicity, and stewardship

This departs from externally imposed environmentalism and makes it a Make-in-India, Decarbonize-for-India mission.

4. Emphasize Economic Security and Global Competitiveness

Rather than couching it in abstract ideals, link the eCarbon Card to:

- Reducing fossil fuel dependency
- Strengthening India's position in future green trade frameworks (e.g., EU's Carbon Border Adjustment Mechanism)
- Creating green micro-economies in rural and urban India

This framing appeals across the political spectrum — to growth-focused policymakers, innovation-driven industries, and sustainability advocates alike.

5. Depoliticize Through Localization

Avoid polarizing language. Instead, ground the system in trusted, grassroots institutions:

- Self-Help Groups, cooperatives, religious institutions, youth clubs, and schools
- Use language of self-reliance, opportunity, and reward
- Demonstrate tangible local benefits (e.g., a village using credits to fund solar lighting)

When the eCarbon Card is embedded in community practice rather than policy discourse, it becomes resilient to political cycles.

6. Leverage Behavioral Psychology & Gamification

People respond best when systems:

- Feel familiar (e.g., rewards, tokens, cashback models like Paytm or Swiggy)
- Are fun and social — badges, rankings, green hero status
- Offer visible progress toward personal and communal goals

This allows even modest behavioral shifts to feel meaningful and validated.

Conclusion:

In an era of politicized sustainability, the eCarbon Card offers a non-partisan, bottom-up pathway to climate action. It focuses on incentives, inclusion, innovation, and economic opportunity — principles that appeal across ideologies. Its potential lies in quietly transforming behaviors at scale, through trust, design, and alignment with national interest.

The eCarbon Card isn't a policy doctrine. It's a platform — one that adapts to context, rewards participation, and turns climate action into a shared, aspirational pursuit rather than a political debate.

Glossary of Key Terms

A
Aadhaar
Aadhaar is the world's largest biometric-based digital identity system, implemented by the Unique Identification Authority of India (UIDAI). It assigns a 12-digit unique identification number to residents, linking demographic and biometric data such as fingerprints and iris scans. The system enables secure authentication and is widely used for government subsidies, banking, mobile connections, and now, potential integration with carbon tracking under the eCarbon Card System.

Accountability Mechanism
A structured system or a framework within the eCarbon Card that ensures individuals, businesses, and institutions are responsible for their carbon emissions. This mechanism includes monitoring, reporting, verification (MRV), and compliance checks.

Adaptability in Carbon Policies
The ability of the eCarbon Card system to evolve over time by adjusting allowances, credit values, and incentives based on technological advancements, climate goals, and policy changes.

Adaptive Carbon Pricing Mechanism

A flexible approach to setting carbon prices that adjusts dynamically based on market conditions, emissions levels, or climate policy objectives. A pricing model that adjusts dynamically based on economic activity, inflation, emissions data, and international market trends, ensuring fair pricing while discouraging excessive emissions.

Additionality (Carbon Credits)

A principle ensuring that carbon credits represent emissions reductions that would not have occurred without the specific intervention or project.

Afforestation

The process of creating new forests on land that was previously not forested to increase carbon sequestration and offset CO_2 emissions.

Allowances (Carbon Allowances)

The amount of CO_2 emissions permitted for an individual or organization under a carbon management system. These allowances can be fixed or adjusted based on policies.

Amazon Web Services (AWS)

A comprehensive and widely adopted cloud computing platform offered by Amazon, providing on-demand services such as data storage, computing power, analytics, and machine learning tools. AWS enables scalable digital infrastructure for organizations and is often used to host applications, manage big data, and support smart systems — including those related to carbon tracking, blockchain, and climate tech solutions.

Annual Carbon Allowance

The maximum amount of CO_2 an individual or business is permitted to emit within a year under the eCarbon Card system. It is adjusted based on sector, demographic, or economic conditions.

Anthropogenic Emissions

Greenhouse gas emissions that result from human activities such as fossil fuel combustion, deforestation, and industrial processes.

Artificial Intelligence for Carbon Tracking (AI-CT)

The use of machine learning and AI algorithms to automate carbon footprint calculations, optimize carbon reduction strategies, and predict emission trends.

Asset-Level Carbon Intensity (ALCI)

A metric, used to measure the carbon footprint of specific financial or physical assets, such as a manufacturing plant, data center, or transport fleet.

Avoided Emissions

Emissions reductions achieved by replacing high-carbon products or services with low-carbon alternatives, such as transitioning from coal-based electricity to solar energy.

B
Baseline Carbon Emissions

The starting point or reference level for measuring carbon reductions. This baseline is typically set based on historical emissions data and used to track progress over time.

Bharat Cloud

An emerging Indian cloud computing initiative aimed at providing indigenous, secure, and scalable digital infrastructure for data storage, processing, and application deployment. Bharat Cloud is envisioned as a sovereign alternative to foreign cloud services, supporting digital public goods, government platforms like Aadhaar and UPI, and national programs such as Digital India. It holds potential to underpin large-scale systems like the eCarbon Card by enabling localized data management, compliance, and real-time analytics within India's regulatory framework.

BharatNet, also known as Bharat Broadband Network Limited (BBNL) - India

BharatNet is an Indian central public sector undertaking, set up by the Department of Telecommunications, a department under the Ministry of Communications of the Government of India for the establishment, management, and operation of the National Optical Fiber Network to provide a minimum of 100 Mbit/s broadband connectivity to all 250,000-gram panchayats in the country, covering nearly 625,000 villages, by improving the middle layer of nation-wide broadband internet in India to achieve the goal of Digital India. It is the world's largest rural broadband connectivity program, which is built under the Make in India initiative with no involvement of foreign companies, It is both an enabler and a beneficiary of other key government schemes, such as Digital India, Make in India, the National e-Governance Plan, UMANG, Bharatmala, Sagarmala, Parvatmala, the dedicated freight corridors, industrial corridors, UDAN-RCS and Amrit Bharat Station Scheme.

Behavioral Economics in Decarbonization

The study of psychological and economic factors that influence individual and corporate decision-making in reducing carbon footprints. It is crucial for designing incentive mechanisms in the eCarbon Card system.

Blockchain for Carbon Transactions

A decentralized, immutable digital ledger that records all carbon credit or carbon related transactions securely and transparently, reducing fraud and ensuring accountability in the carbon market.

Blue Carbon

Carbon stored in coastal and marine ecosystems such as mangroves, seagrasses, and salt marshes, which play a crucial role in climate mitigation.

Border Carbon Adjustments (BCA)

A tax or tariff imposed on imported goods based on their carbon footprint to prevent "carbon leakage" and ensure global competitiveness in carbon pricing.

Business-as-Usual (BAU) Scenario

A projection of future carbon emissions assuming no additional climate policies or mitigation efforts beyond what is currently implemented.

C

California Climate Credit

The California Climate Credit is a program established by the California Public Utilities Commission (CPUC) as part of the state's strategy to reduce greenhouse gas emissions under its cap-and-trade system. Funded by revenues from California's carbon pollution pricing program, it returns money directly to consumers to encourage sustainable choices and offset utility costs.

Cap-and-Trade System - California

A regulatory framework that sets a cap on total carbon emissions and allows entities to trade carbon credits to stay within their allocated limits. A market-based approach to controlling pollution by providing economic incentives for reducing emissions. A cap is set on total emissions, and businesses can trade unused allowances.

Carbon Accounting

The systematic tracking, measuring, and reporting of carbon emissions across various activities, sectors, and industries.

Carbon Authority of India (CAI)

A proposed (hypothetical) government regulatory body responsible for overseeing the implementation and governance — monitoring, and enforcement of the eCarbon Card system in India.

Carbon Budget

The maximum amount of CO_2 that can be emitted globally while keeping global temperature rise within a specific limit, such as 1.5°C or 2°C.

Carbon Capture and Storage (CCS)

A technology that captures carbon emissions from industrial processes and stores them underground to prevent atmospheric release.

Carbon Capture, Utilization, and Storage (CCUS)

Carbon Capture, Utilization, and Storage (CCUS) refers to a set of emerging technologies designed to capture carbon dioxide (CO_2) emissions from power plants, industrial facilities, or directly from the atmosphere, and either reuse it or store it permanently underground to prevent it from entering the atmosphere.

Carbon Credits

Tradable permits representing one metric ton of CO_2-equivalent reduction, which can be earned by reducing emissions or supporting green projects, bought and sold in carbon markets.

Carbon Credit Card

A digital or physical card that tracks an individual's or organization's carbon emissions and facilitates the earning, trading, or redemption of carbon credits based on their environmental behavior. It links personal or institutional activities — such as transportation, energy use, or consumption — to a carbon budget, enabling users to stay within emission limits, purchase additional credits, or earn rewards for low-carbon choices. The Carbon Credit Card is a key tool for democratizing climate accountability and integrating market-based incentives into daily life.

Climate Credit Card (Switzerland)

A pilot initiative launched in Switzerland that allows users to track and understand the carbon footprint of their purchases in real time. Developed through collaborations between Swiss fintechs, climate researchers, and sustainability organizations, the card links spending data to estimated CO_2 emissions and offers behavioral nudges, insights, and offset options. It exemplifies how financial tools can be leveraged to drive individual climate action and informed consumption, serving as an early model for personalized carbon accountability systems like the eCarbon Card.

Carbon Disclosure Project (CDP)

CDP (formerly Carbon Disclosure Project) is a global non-profit organization that runs the world's leading environmental disclosure system for companies, cities, states, and regions. CDP enables entities to measure, manage, and disclose their environmental data, particularly on climate change, water security, and deforestation — and helps them reduce their carbon footprints in alignment with global climate goals.

Carbon Footprint

The total amount of greenhouse gases emitted directly and indirectly by an individual, company, or product, measured in CO_2-equivalent (CO_2e).

Carbon Intensity

A measure of the amount of CO_2 emitted per unit of economic output, energy produced, or product manufactured. Typically, it is the amount of CO_2 emissions per unit of economic output (e.g., GDP per ton of CO_2), often used as an indicator of a country's or company's environmental efficiency.

Carbon Labeling on Consumer Goods
A system in which products display their estimated carbon footprint, allowing consumers to make more informed and sustainable purchasing choices.

Carbon Leakage
The transfer or relocation of carbon-intensive industries from one country to countries with less stringent emissions regulations, potentially offsetting overall climate gains thus overall leading to no net reduction in global emissions.

Carbon Literacy Training
Educational programs, designed to increase awareness of carbon footprints, emissions reduction strategies, and sustainable decision-making.

Carbon Marketplace
A trading platform where individuals and businesses buy and sell carbon credits to meet their emission reduction goals.

Carbon Microtransactions
A mechanism where very small carbon credits (fractions of a ton) can be exchanged, allowing individuals to participate in carbon markets with everyday activities.

"Carbon-plus" Economy

A "carbon-plus" economy, or a carbon-positive economy, refers to an economic system that not only reduces carbon emissions but also captures and stores more carbon than it releases. This means moving beyond simply limiting emissions to actively removing carbon from the atmosphere and enhancing natural carbon sinks, such as forests and oceans.

Carbon Neutrality

A state in which an entity balances its carbon emissions with carbon removal measures, achieving net-zero emissions.

Carbon Offsetting

A process where organizations or individuals invest in projects that reduce or remove carbon emissions, such as renewable energy or reforestation, to compensate for their own emissions.

Carbon Pricing

An economic policy tool used to place a monetary value on carbon emissions i.e. assigning cost to emitting CO_2, either through a carbon tax or a cap-and-trade system, to incentivize emission reductions.

Carbon Quotas

Pre-determined emission limits assigned to individuals or corporations, which can be adjusted over time as climate goals evolve.

Carbon Sink

Natural systems such as forests, soil, and oceans that absorb and store more carbon dioxide than they release.

Carbon Tokenization

The process of converting carbon credits into digital tokens that can be traded on blockchain platforms, enhancing transparency and accessibility.

Circular Economy

A regenerative economic system that minimizes waste by designing products for reuse, repair, and recycling, reducing overall carbon emissions.

Climate Account (Klimatkontot)

The Climate Account (Klimatkontot) is a personal carbon footprint tracker used in Sweden, allowing individuals to monitor their emissions based on daily activities, travel, and energy consumption. It provides a self-assessment tool for reducing carbon footprints, which aligns with the eCarbon Card's personal tracking goals.

Climate Calculation Engine

A Climate Calculation Engine is a data-driven algorithmic system that computes individual, corporate, or national carbon footprints based on real-time consumption, energy usage, and emission factors. Such engines power Personal Carbon Wallets, carbon allowances, and tax policies.

Climate Equity Adjustment (CEA)

A policy approach that considers social and economic inequalities when designing carbon pricing and emission reduction strategies.

Climate Leap Initiative (Klimatklivet) - Sweden

Sweden's Climate Leap (Klimatklivet) is a public funding program that supports local climate projects such as electrification of transport, bioenergy, and energy efficiency upgrades. This initiative provides a blueprint for implementing localized carbon reduction projects under the eCarbon framework.

Climate Neutrality

A broader concept than carbon neutrality, covering all greenhouse gases and not just CO_2 emissions. Achieving net-zero greenhouse gas emissions by balancing emissions with carbon removal through offsetting or sustainable practices.

Climate Resilience

The ability of an individual, community, business, or ecosystem to anticipate, prepare for, and respond to climate-related disruptions or climate change impacts.

Consumption-Based Emissions Accounting

A method of calculating emissions based on what individuals or businesses consume rather than where production occurs.

The Council on Energy, Environment and Water (CEEW)

It is one of India's leading independent public policy think tanks, known for rigorous, data-driven research and deep engagement on issues related to energy transition, climate change, and sustainable development.

Cumulative Emissions Budget

A fixed amount of CO_2 that can be emitted globally or nationally over a specified period to remain within climate targets.

D

Decarbonization

The process of reducing carbon dioxide emissions across all sectors, including energy, transportation, and industry through sustainable practices and clean technologies.

Deforestation Emissions

CO_2 emissions resulting from the clearing of forests for agriculture, logging, or development, which reduces natural carbon sinks.

Department for Environment, Food & Rural Affairs (DEFRA) – UK

DEFRA is the UK government department responsible for environmental protection, food production and standards, agriculture, fisheries, and rural communities. It plays a central role in the UK's climate policy and carbon accounting infrastructure, especially in shaping how emissions are measured, reported, and mitigated at both the public and private levels.

DigiLocker – India

DigiLocker is a flagship digital initiative by the Ministry of Electronics and Information Technology (MeitY), Government of India, that allows citizens to store, access, and share official documents in a secure, cloud-based digital locker. It plays a pivotal role in India's digital governance ecosystem, enabling paperless services, real-time document verification, and e-governance efficiency.

Digital Carbon Tracking

The use of technology, such as IoT devices, blockchain, and AI, to measure, monitor and record carbon emissions from individual and corporate activities in real-time.

Digital Peer-to-Peer (P2P) Marketplace

An online platform that enables individuals or small entities to directly exchange goods, services, or digital assets — such as carbon credits — without traditional intermediaries. In the context of the eCarbon Card system, a digital P2P marketplace facilitates the transparent and real-time trading of surplus or deficit carbon allowances between users, promoting decentralized participation in climate action. Often powered by blockchain or smart contracts, such platforms ensure secure, auditable, and equitable transactions across a wide user base.

Digital Public Infrastructure (DPI)

Foundational technology platforms — such as digital identity, payment systems, and data exchange protocols — that enable secure, inclusive, and scalable public service delivery. Examples in India include Aadhaar (identity), UPI (payments), DigiLocker (digital records), and ONDC (commerce). DPI plays a critical role in enabling large-scale systems like the eCarbon Card by providing the backbone for real-time verification, transaction tracking, and equitable access to digital services across the population.

Distributed Ledger for Carbon Tracking

A blockchain-enabled system that records and verifies emissions transactions in a transparent, tamper-proof manner.

Distributed Renewable Energy (DRE)

Decentralized energy systems that generate electricity from renewable sources at the point of consumption, reducing transmission losses and emissions.

Dynamic Allowance Adjustments

The practice of periodically recalibrating individual or corporate carbon allowances based on progress in emissions reduction and economic growth.

Dynamic Carbon Allowance System

An emissions cap mechanism that adjusts in real-time, based on changes in national emissions intensity, economic fluctuations, and technological advancements.

E

eCarbon Card

A digital carbon management system that assigns carbon allowances to individuals and businesses, tracks emissions, and enables carbon credit trading through financial transactions.

e-Residency - Estonia

Estonia's e-Residency program allows non-residents to digitally establish and manage businesses within the country while accessing Estonia's digital governance system. It provides a secure digital identity, enabling e-business, banking, and taxation without physical presence. Similar digital residency models could be leveraged to develop a cross-border carbon trading network within the eCarbon framework.

e-Governance (Estonia)

A globally recognized digital governance model pioneered by Estonia that enables nearly all public services — including voting, taxation, healthcare, and business registration — to be accessed securely online. Powered by a foundational digital ID system and blockchain-backed infrastructure, Estonia's e-Governance system demonstrates how digital trust, transparency, and efficiency can be scaled nationally. It serves as a valuable reference for implementing secure, citizen-centric platforms like the eCarbon Card system in other countries.

Eco-Points Program – Japan[31]

Eco Points is a pioneering green incentive program launched by the Government of Japan to encourage consumers to adopt environmentally friendly practices—primarily by purchasing energy-efficient appliances. It offers a real-world example of reward-based decarbonization at the citizen level, highly relevant to the logic of the eCarbon Card system.

Emission Factors

Standardized values used to estimate the amount of greenhouse gases emitted per unit of activity, such as per liter of fuel consumed or per kilowatt-hour of electricity used.

Emission Reduction Targets

Specific goals, set by governments or corporations to lower carbon or greenhouse gas emissions over a set period.

Emission Trading Scheme (ETS) - South Korea

Launched in 2015, the South Korea ETS is the first nationwide cap-and-trade system in East Asia, covering major industrial sectors such as power generation, steel, petrochemicals, and aviation. It sets emissions caps for large emitters and allows the trading of allowances to incentivize cost-effective reductions. The system incorporates auctioning, benchmarking, and offset credits. As one of the most comprehensive carbon markets in Asia, it provides a valuable precedent for developing carbon credit systems like the eCarbon Card — especially in terms of policy design, compliance mechanisms, and market dynamics.

European Union Emission Trading System (EU ETS)

The EU ETS is the world's largest carbon trading system, setting legally binding carbon caps for industries and power plants. Companies exceeding their emissions must purchase additional carbon credits, while those under their limit can sell excess allowances. The eCarbon Card System aims to adapt similar market-driven carbon control at the individual and business levels.

United Kingdom Emission Trading System (UK ETS)

The UK implemented its own emissions trading scheme (UK ETS) on January 1, 2021, after leaving the European Union, which is a cap-and-trade system, but for businesses rather than individuals. The cap is reduced in line with the UK's 2050 net-zero commitment.

Energy Efficiency Incentives

Financial benefits, such as tax breaks or subsidies, for businesses and individuals that adopt energy-saving technologies.

Environmental, Social, and Governance (ESG) Criteria

A set of non-financial performance indicators that assess a company's commitment to sustainability, social responsibility, and ethical governance.

European Green Deal

The European Green Deal is the European Union's flagship strategy to become the world's first climate-neutral continent by 2050. Launched in December 2019, it lays out a comprehensive roadmap for transforming the EU's economy and society to meet environmental and climate goals—without leaving anyone behind. It represents not just an environmental policy, but an economic, social, and industrial transformation agenda backed by regulation, finance, innovation, and diplomacy.

European Carbon Allowances (EUA)

Tradable permits issued under the European Union Emissions Trading System (EU ETS), representing the right to emit one metric ton of CO_2 or its equivalent in other greenhouse gases. EUAs are the primary units of compliance in the EU's cap-and-trade system, where a total emissions cap is set, and companies can buy or sell allowances based on their emissions performance. The EUA market sets a transparent price for carbon, serving as a global benchmark and informing pricing mechanisms in emerging carbon trading systems like the eCarbon Card.

F
FPO (Farmer Producer Organization)

A legally registered collective of farmers, typically organized as a cooperative or producer company, aimed at improving the livelihoods and bargaining power of small and marginal farmers. FPOs enable members to collectively procure inputs, access credit, adopt technology, and market their produce more efficiently. In the context of the eCarbon Card system, FPOs can serve as crucial intermediaries for aggregating low-carbon agricultural practices, managing carbon credit earnings, and facilitating climate-smart farming at scale.

G
Gamification

The application of game-design elements — such as points, levels, badges, challenges, and leaderboards — in non-game contexts to motivate engagement, influence behavior, and enhance user experience. In the context of the eCarbon Card system, gamification can encourage sustainable practices by rewarding individuals or communities for reducing emissions, staying within carbon allowances, or participating in green initiatives, thereby making climate action more engaging, social, and rewarding.

Geographical Information System (GIS)

A GIS is a spatial data analysis tool used for mapping and monitoring environmental factors, emissions hotspots, and land-use changes. In the eCarbon framework, GIS supports tracking urban pollution, deforestation, and renewable energy distribution.

Google Pay

A widely used digital payment platform developed by Google that enables secure, real-time transactions using smartphones or other internet-connected devices. In India, Google Pay operates on the Unified Payments Interface (UPI) system, allowing users to send money, pay bills, make purchases, and access financial services with ease. Its integration with India's Digital Public Infrastructure makes it a potential interface for tools like the eCarbon Card — supporting carbon tracking, incentives, and peer-to-peer carbon credit trading through familiar payment ecosystems.

The Goods and Services Tax System (GST)

India's GST is a unified indirect tax system, streamlining taxation across states. A similar framework could be used to implement Carbon Taxation and Personal Carbon Credits, ensuring nationwide integration of decarbonization efforts.

Green Bonds

Financial instruments that raise capital for climate-related and environmental projects, often offering tax incentives or lower interest rates for sustainability efforts.

GCF (Green Climate Fund)

An international financial mechanism established under the United Nations Framework Convention on Climate Change (UNFCCC) to support developing countries in responding to climate change. The GCF finances low-emission and climate-resilient projects, including renewable energy, sustainable agriculture, and adaptation infrastructure. For initiatives like the eCarbon Card, the GCF can be a potential funding partner — supporting pilot programs, technological integration, and capacity-building efforts aligned with national climate goals.

Green Infrastructure

Public and private investments in energy-efficient or sustainable buildings, smart or efficient transport systems or other eco-friendly infrastructure, and sustainable urban development.

Greenwashing

A deceptive practice where companies exaggerate or falsely claim their sustainability efforts to appeal to environmentally conscious consumers.

Greenhouse Gases (GHGs)

GHGs are gases in the Earth's atmosphere that trap heat, contributing to the greenhouse effect and global warming. While they occur naturally, human activities — especially since the Industrial Revolution — have dramatically increased their concentration, leading to climate change.

Green Rewards – United Arab Emirates (UAE)

Green Rewards is a digital incentive program launched by the UAE government to encourage environmentally responsible behavior among citizens and residents by rewarding sustainable lifestyle choices. It reflects the UAE's growing commitment to climate action, digital innovation, and public engagement in its Net Zero by 2050 strategy.

Gross Domestic Product (GDP) per Emission Unit

A ratio that measures a country's economic output relative to its carbon emissions. An economic indicator showing the relationship between a country's economic output and its carbon emissions.

I

IoT-Enabled Emission Monitoring

The use of Internet of Things (IoT) sensors to track and record carbon emissions from energy use, transportation, and industrial processes.

Incentivization Model

A system of financial rewards and penalties, designed to encourage sustainable behavior among individuals and corporations.

India Greenhouse Gas Program (India GHG Program)

A voluntary, industry-led initiative launched by WRI India, CII, and TERI to build comprehensive greenhouse gas (GHG) measurement and management capacity among Indian businesses. The program provides technical guidance, tools, and platforms to help organizations measure their emissions, set reduction targets, and disclose progress transparently. It plays a key role in fostering a culture of carbon accountability and complements national decarbonization efforts — including potential integration with systems like the eCarbon Card for corporate-level emissions tracking.

International Energy Agency (IEA)

A global authority on energy policy, data, and modeling, providing rigorous analysis and actionable recommendations on clean energy transitions.

Intergovernmental Panel on Climate Change (IPCC)

The world's leading scientific body for assessing climate change. Established by the United Nations Environment Program (UNEP) and the World Meteorological Organization (WMO) in 1988, the IPCC does not conduct original research, but provides comprehensive, consensus-based assessments of scientific, technical, and socio-economic information related to climate change.

International Solar Alliance (ISA) – India, France

Founded by India and France, the International Solar Alliance (ISA) is a global treaty-based organization promoting solar energy deployment in tropical countries. ISA advances solar financing, technology transfer, and capacity-building, which could play a role in the renewable energy incentives of the eCarbon system.

K

Klimmatpontol (Sweden's Climate Account Tool)

A publicly available carbon footprint calculator created by Uppsala University, SEI (Stockholm Environment Institute), and other Swedish climate organizations. It is designed to help individuals estimate their annual CO_2e emissions based on key lifestyle areas: housing and energy use, transportation (car, flights, public transit), food consumption patterns and shopping and general consumption

M

Ministry of Environment, Forest and Climate Change (MoEFCC) - India

The nodal government ministry in India responsible for planning, promoting, coordinating, and overseeing the implementation of environmental and climate change policies.

Mahatma Gandhi National Rural Employment Guarantee Act (MGNREGA) - India

The **Mahatma Gandhi National Rural Employment Guarantee Act (MGNREGA)**, enacted in 2005, is one of India's most ambitious and transformative rural welfare programs. It provides guaranteed wage employment to rural households while building community assets and promoting sustainable development.

MSME (Micro, Small and Medium Enterprises

A vital segment of India's economy, MSMEs encompass businesses classified by investment and turnover thresholds, ranging from small shops and service providers to manufacturing units. They contribute significantly to employment, exports, and grassroots innovation. In the context of the eCarbon Card system, MSMEs play a key role in decentralized carbon management — both as emitters and as potential beneficiaries of low-carbon incentives, technology upgrades, and carbon credit trading opportunities.

Ministry of New and Renewable Energy (MNRE) – India

Responsible for research and development, intellectual property protection, and international cooperation, promotion, and coordination in renewable energy sources such as wind power, small hydro, Battery Energy Storage and solar power.

Ministry of Housing and Urban Affairs (MoHUA) – India

Responsible for the formulation and administration of the rules and regulations and laws relating to the housing and urban development in India.

Mission LiFE – Lifestyle for Environment – India

Mission LiFE (Lifestyle for Environment) is a flagship global initiative launched by Prime Minister Narendra Modi in partnership with the United Nations, aimed at transforming individual and community behavior to address climate change. It encourages people to adopt environmentally responsible lifestyles, making "pro-planet behavior" the norm instead of the exception.

Mitigation Strategies

Actions taken to reduce or prevent greenhouse gas emissions, such as improving energy efficiency, transitioning to renewable energy or clean technology adoption.

Mukhya Mantri Solar Yojana – India

The Mukhya Mantri Solar Yojana is a state-level initiative launched by the Delhi government to promote rooftop solar energy adoption among residents, particularly in group housing societies and low-income households. It aligns with India's broader clean energy targets while helping decentralize solar power generation within the national capital.

My Tree Campaign – India

"My Tree Campaign" (also known as Nanna Mara in Kannada) is an urban greening initiative launched by the Bengaluru City Administration to promote citizen-led afforestation, urban biodiversity, and environmental stewardship in the rapidly growing metropolis. The campaign empowers residents to plant, adopt, and care for trees, creating a sense of ownership and participation in the city's environmental well-being.

N

National Carbon Data Hub

India's National Carbon Data Hub is envisioned as a centralized digital repository for monitoring and analyzing national and sectoral carbon emissions. The hub will integrate satellite data, corporate reports, and IoT-based carbon tracking to inform climate policies and support eCarbon trading.

National Carbon Market – China

China's National Carbon Market is the largest emissions trading system (ETS) in the world, launched in July 2021 to regulate and reduce industrial carbon emissions. It operates under a cap-and-trade system, where companies receive carbon allowances and must either reduce emissions or purchase extra credits if they exceed their limits. Initially covering power sector emissions, the market plans to expand to industrial, transportation, and petrochemical sectors. This system supports China's goal of carbon neutrality by 2060 and is a key component of its climate strategy. The market is expected to introduce auction-based allowance allocation, link with regional pilot programs, and align with international carbon markets. The eCarbon Card System can adopt similar cap-and-trade principles for personal and corporate carbon trading, offering market-driven incentives for reduction of emissions.

NIC Cloud (National Informatics Centre Cloud)

A government-managed cloud computing platform developed by the National Informatics Centre (NIC) under India's Ministry of Electronics and Information Technology. Also known as MeghRaj, the NIC Cloud provides secure, scalable, and cost-effective infrastructure for hosting government applications, services, and digital public goods. It supports mission-critical platforms like Aadhaar and DigiLocker and could serve as a foundational backend for implementing systems like the eCarbon Card — ensuring sovereign data control, privacy, and interoperability across departments.

National Solar Mission - India

India's National Solar Mission (NSM) is a key pillar of its climate policy, aiming for 280 GW of solar energy by 2030. The mission promotes solar subsidies, solar parks, and rooftop installations, aligning with eCarbon Card's renewable energy incentives.

Net-Zero Emissions

A balance between the amount of greenhouse gases emitted and the amount removed from the atmosphere, typically achieved through carbon offsets or sequestration.

Nationally Determined Contributions (NDCs)[25]

Climate action plans submitted by countries under the Paris Agreement1 outlining their commitments to reducing green-house gas emissions.

National Institution for Transforming India (NITI) Aayog – India

Serves as the apex public policy think tank of India and the nodal agency tasked with catalyzing economic development and fostering cooperative federalism and moving away from bargaining federalism through the involvement of State Governments of India in the economic policy-making process using a bottom-up approach.

National Sample Survey Office (NSSO) – India

A part of the Ministry of Statistics and Program Implementation in India, responsible for conducting large-scale sample surveys on various socio-economic topics

Offline Kiosks – A Critical Infrastructure for eCarbon Card Inclusion

To ensure universal access and digital equity, especially in rural and low-connectivity areas, the deployment of offline eCarbon Card Kiosks is a vital part of the system's outreach and operational architecture. These kiosks serve as physical access points for individuals and small businesses to engage with the eCarbon

Card ecosystem, even if they do not own smartphones or have internet access.

ONDC – Open Network for Digital Commerce (India)

ONDC (Open Network for Digital Commerce) is a groundbreaking initiative by the Government of India, launched under the aegis of the Department for Promotion of Industry and Internal Trade (DPIIT). It aims to democratize digital commerce by creating a platform-agnostic, open-source network where buyers and sellers can transact directly, regardless of which e-commerce platform they use. It's often referred to as the "UPI for e-commerce", as it brings interoperability, transparency, and inclusivity to India's digital marketplace.

One Nation, One Ration Card (ONORC) – India

ONORC is an Indian government initiative enabling ration card portability across states. It ensures food security by allowing beneficiaries to access subsidized food grains anywhere in the country using their Aadhaar-linked ration card. The model could inspire the national portability of personal carbon allowances, ensuring fair and inclusive distribution.

P
PAN Card – India

The Permanent Account Number (PAN) is a 10-character alphanumeric identifier issued by the Indian Income Tax Department. It is used primarily for taxation purposes, financial transactions, and regulatory compliance. PAN card linkage with carbon allowances and personal carbon credits could provide an effective mechanism for tracking high-carbon financial transactions.

Paytm

One of India's leading digital payment and financial services platforms, Paytm enables users to make cashless transactions, pay bills, transfer money, and access banking and insurance services via mobile devices. Operating on the Unified Payments Interface (UPI), it plays a major role in promoting financial inclusion and digital commerce. In the context of the eCarbon Card system, platforms like Paytm could serve as interfaces for carbon credit transactions, behavioral incentives, and integration with digital public infrastructure.

Panchayat

A grassroots unit of local self-government in rural India, operating at the village, intermediate, and district levels under the Panchayati Raj system. Panchayats are responsible for planning and implementing development programs, managing local resources, and representing community interests. In the context of the eCarbon Card system, Panchayats can play a critical role in community outreach, carbon literacy, distribution of carbon allowances, and facilitation of low-carbon initiatives in rural areas.

Paris Agreement

An international treaty signed in 2015 that aims to limit global temperature rise to well below 2°C above pre-industrial levels, with efforts to keep it at 1.5°C.

Perform, Achieve, Trade (PAT) – India

A market-based energy efficiency trading mechanism launched by the Bureau of Energy Efficiency (BEE) under India's

National Mission on Enhanced Energy Efficiency (NMEEE). The PAT scheme sets specific energy reduction targets for large industries and allows them to trade Energy Saving Certificates (ESCerts) if they exceed or fall short of their targets. It incentivizes cost-effective energy savings and emissions reductions in energy-intensive sectors. The PAT framework serves as a key precedent for developing broader carbon credit systems like the eCarbon Card.

Personal Carbon Allowance – UK

A conceptual framework proposed in the early 2000s by the UK government and researchers to allocate individuals a fixed annual carbon budget for activities like transport, home energy use, and consumption. Each person would track and manage their emissions, with the ability to trade unused allowances. Though never implemented at scale, the UK's exploration of Personal Carbon Allowances (PCAs) laid important groundwork for citizen-centric climate accountability systems like the eCarbon Card.

Personal Carbon Wallet

A Personal Carbon Wallet is a digital tool that tracks an individual's carbon allowance, expenditures, and credits. Users can monitor their carbon footprint, purchase additional carbon credits, and redeem incentives for low-emission behaviors. This is a core concept of the eCarbon Card System.

Personal Carbon Trading (PCT)

Personal carbon trading, is the generic term for a number of proposed carbon emissions trading schemes under which emissions credits would be allocated to adult individuals on a (broadly) equal per capita basis, within national carbon budgets. Individuals then surrender these credits when buying fuel or electricity, for example. Individuals who exceed their allocation (i.e. those who want to use more emissions credits than they have been given or permitted by their initial allocation) would be able to purchase additional credits from those who emit less than their initial allocation, so individuals that are under allocation would profit from their small carbon footprint. Some forms of personal carbon trading (carbon rationing) could be an effective component of climate change mitigation. Research in this area has shown that personal carbon trading would be a progressive policy instrument — redistributing money from the rich to the poor — as the rich use more energy than the poor and so would need to buy allowances from them. This is in contrast to a direct carbon tax, under which all lower income people are worse off, prior to revenue redistribution.

Personal Carbon Trading System (PCT) – UK

A type of cap-and-trade scheme was explored in the UK but not implemented as a national policy, where individuals would receive tradable carbon allowances for emissions from household energy and/or personal travel.

Personally Identifiable Information (PII)

Also known as personal data includes information that relates to an identified or identifiable natural person, meaning someone can be identified directly or indirectly through that information.

Personalized Carbon Allowance (PCA)

An individualized carbon budget assigned based on lifestyle, region, and consumption patterns to ensure fairness in emissions reductions.

PhonePe - India

A widely used Indian digital payments platform built on the Unified Payments Interface (UPI), PhonePe allows users to make instant money transfers, pay utility bills, shop online, and invest in financial products. Known for its user-friendly interface and deep penetration in both urban and rural markets, PhonePe plays a key role in India's digital economy. Within the eCarbon Card ecosystem, platforms like PhonePe could enable seamless carbon credit purchases, rewards disbursal, and real-time emissions tracking through integrated transactions.

Price Ceiling (Carbon Pricing Ceiling)

The maximum allowable price for a carbon credit in a regulated carbon trading system. A price ceiling prevents carbon credits from becoming prohibitively expensive, ensuring affordability and market stability — especially for low-income participants or essential sectors. In the eCarbon Card system, a price ceiling helps protect users from extreme cost spikes during high demand, while balancing economic feasibility with environmental impact.

Price Floor (Carbon Pricing Floor)

The minimum allowable price set for a carbon credit within a trading or carbon pricing system. A floor ensures that carbon credits do not become too cheap, maintaining a meaningful economic incentive for emission reductions. In the eCarbon Card system, a price floor prevents devaluation of surplus credits and guarantees that low-carbon behaviors retain financial value for individuals and businesses participating in carbon trading.

PM KUSUM (Pradhan Mantri Kisan Urja Suraksha evam Utthaan Mahabhiyan) – India

Launched in 2019 by the Ministry of New and Renewable Energy (MNRE), PM-KUSUM is a government-backed initiative to promote solar energy adoption among farmers — reducing dependence on grid electricity and diesel, while enhancing farmer income and energy security.

PM Ujjwala (LPG) – Pradhan Mantri Ujjwala Yojana – India

Launched in May 2016, PM Ujjwala Yojana (PMUY) is a flagship welfare scheme of the Government of India aimed at providing clean cooking fuel (LPG) to women from poor households, especially in rural and underprivileged areas. It is a crucial initiative for improving health, gender equity, and environmental outcomes.

Public Distribution System (PDS) – India

The PDS is a government-run distribution network that provides subsidized food grains and other essential items to eligible households through Fair Price Shops (FPS), commonly known as ration shops. It is implemented under two major schemes: (I) National Food Security Act (NFSA), 2013 and (II) Targeted Public Distribution System (TPDS)

R

Regenerative Agriculture for Carbon Sequestration

Farming practices that restore soil carbon, increase biodiversity, and improve water retention, enhancing natural carbon sinks.

Renewable Energy Certificates (RECs)

Market-based instruments that certify that a certain amount of electricity was generated from renewable sources, allowing businesses to offset their energy-related emissions.

The Royal Commission on Environmental Pollution (RCEP) – UK

It was a UK environmental advisory body established in 1970 by Royal Warrant to advise the Queen, Government, Parliament, and the public on environmental issues, and was abolished in 2011.

The Royal Society for the Encouragement of Arts, Manufactures and Commerce, commonly known as the Royal Society of Arts (RSA) – UK

An organization based in London, UK that champions innovation and progress across a multitude of sectors by fostering creativity, social progress, and sustainable development. Through its extensive network of changemakers, thought leadership, and projects, the RSA seeks to drive transformative change, enabling "people, places, and the planet to thrive in harmony."

S

Saubhagya Scheme (*aka* Pradhan Mantri Sahaj Bijli Har Ghar Yojana) – India

The Saubhagya Scheme is an Indian government initiative aimed at universal household electrification, by providing electricity connections to all un-electrified households in rural areas and all poor households in urban areas in the country. This scheme provides an integration pathway for carbon accounting in rural energy use.

Self-Determination Theory (SDT)

A psychological framework that explains human motivation based on the fulfillment of three innate needs: autonomy, competence, and relatedness. When these needs are met, individuals are more likely to adopt and sustain positive behaviors. In the context of the eCarbon Card system, SDT supports the design of incentives, feedback mechanisms, and engagement strategies that empower citizens to voluntarily participate in low-carbon behaviors, driven not by coercion but by intrinsic motivation and social belonging.

Self Help Groups (SHGs)

Small, community-based voluntary associations — primarily of women — that pool savings and offer microloans, mutual support, and capacity building at the grassroots level. SHGs play a vital role in financial inclusion, rural development, and social empowerment in India. Within the eCarbon Card system, SHGs can act as local facilitators for carbon literacy, credit aggregation, and behavior change campaigns, especially in rural and underserved areas, enabling inclusive and community-driven climate action.

Supply Chain Emissions

Greenhouse gas emissions that occur throughout the entire supply chain, from raw material extraction to product disposal.

Science Based Targets initiative (SBTi)

A global collaboration between CDP, the UN Global Compact, World Resources Institute (WRI), and WWF that helps companies set greenhouse gas reduction targets aligned with climate science and the goals of the Paris Agreement. SBTi validates whether corporate climate targets are consistent with limiting global warming to well below 2°C, ideally 1.5°C. In the context of the eCarbon Card system, SBTi provides a benchmarking framework for aligning business carbon allowances and reductions with scientifically grounded climate action pathways.

Smart Cities Mission – India

Launched in 2015, India's Smart Cities Mission (SCM) is an urban renewal initiative aimed at developing 100 smart cities through technology-driven governance, sustainable infrastructure, and digital solutions. It focuses on smart mobility, energy efficiency, waste management, and e-governance to enhance urban livability and sustainability. The mission integrates renewable energy, intelligent transportation, and green urban planning, supporting India's climate and decarbonization goals. The eCarbon Card System could complement SCM by enabling personal carbon tracking, incentivizing low-carbon choices, and integrating sustainability metrics into urban services.

Smart Metering for Carbon Tracking

Advanced digital meters that provide real-time data and insights into energy usage and carbon emissions, enabling users to make informed decisions on reducing their carbon footprint.

Smart Nation Initiative

Singapore's Smart Nation Initiative integrates technology and digital governance into daily life, enabling real-time urban management, electronic payments, and AI-driven policymaking. The initiative fosters sustainability through smart grids, mobility solutions, and climate monitoring, providing an ideal digital infrastructure for eCarbon Card deployment.

Small and Medium Enterprises (SMEs)

SMEs are the backbone of India's economy, contributing over 30 percent of GDP and employing more than 110 million people. They span diverse sectors—from textiles and food processing to electronics and auto parts — and operate in both urban and rural contexts.

Sole & Small Micro Enterprises (SSMEs)

SSMEs are one-person or family-run informal businesses such as tailors, cobblers, mobile repairers, and local service providers. Often operating in rural areas or informal urban settlements, they typically fall below the GST threshold and are not formally registered. With minimal mechanization and low — but not negligible — carbon footprints, SSMEs may not meet formal SME definitions but are vital to local economies, livelihoods, and grassroots climate mitigation efforts.

Sustainable Energy Fund – UNDP

A funding mechanism established by the United Nations Development Program (UNDP) to support access to clean, affordable, and sustainable energy in developing countries. The fund mobilizes public and private finance to scale renewable energy projects, promote energy efficiency, and strengthen energy access for underserved populations. It aligns with UNDP's broader goals of achieving the Sustainable Development Goals (SDGs), particularly SDG 7 (Affordable and Clean Energy). In the context of the eCarbon Card system, such funds offer potential pathways for financing pilot programs and incentivizing green transitions in low-carbon communities.

Sustainable Energy Fund for Africa (SEFA)

A multi-donor trust fund managed by the African Development Bank (AfDB) that provides catalytic finance to unlock private investments in renewable energy and energy efficiency across Africa. SEFA supports projects through grants, concessional finance, and technical assistance, focusing on clean energy access, green mini-grids, and innovative climate technologies. Though region-specific, SEFA serves as a model for how blended finance can accelerate energy transitions — offering insights relevant to scaling initiatives like the eCarbon Card in developing economies.

Swachh Bharat Mission – India

The Swachh Bharat Mission (Clean India Mission) is India's nationwide sanitation drive to eliminate open defecation and improve waste management. It offers lessons in behavioral change, which is critical for the mass adoption of personal carbon accountability.

T

Tax Rebates for Carbon Efficiency

Financial incentives, provided to businesses and individuals that stay below their carbon allowances or invest in clean, sustainable technologies and thus reduce their carbon footprint.

The Energy and Resources Institute (TERI) India

TERI is one of India's foremost think tanks and research institutions focused on energy, environment, and sustainable development.

U

Ujjwala Yojana - India

As an Indian government initiative, the Pradhan Mantri Ujjwala Yojana provides subsidized LPG cylinders to low-income households, replacing polluting biomass fuel. This aligns with eCarbon Card incentives for promoting clean energy access.

Unified Payments Interface (UPI) for Carbon Transactions | UPI-Enabled Carbon Transactions – India

UPI is India's real-time digital payments system, allowing seamless transactions between bank accounts using a single mobile application. The eCarbon Card could leverage UPI for carbon credit transactions, integrating sustainability into daily purchases. The integration of carbon tracking and carbon credit trading into digital payment systems to streamline transactions, allow seamless deductions and rewards in daily transaction and enhance accessibility.

United Nations Framework Convention on Climate Change (UNFCCC)

Central international treaty body for coordinating global climate action. Signed in 1992 and ratified by 198 countries (Parties), the UNFCCC provides the legal and diplomatic framework for initiatives like the Paris Agreement1, Nationally Determined Contributions (NDCs)25, carbon markets, adaptation finance, and climate transparency.

Unstructured Supplementary Service Data (USSD)

USSD is a mobile communication protocol used by GSM cellular phones to interact directly with service providers via short codes (e.g., *123#). Unlike SMS, USSD sessions are real-time and do not require internet connectivity. In the context of the eCarbon Card, USSD enables users — especially in rural or low-digital-access areas — to check their carbon balances, redeem rewards, or receive notifications without needing a smartphone or data plan, making the system more inclusive and accessible across socioeconomic segments.

V

Value Chain Decarbonization Strategy

A corporate framework for reducing emissions across all stages of production, from raw material sourcing to end-user consumption.

W

Water-Energy-Carbon Nexus

An interdisciplinary approach that examines the interconnections between water usage, energy production, and carbon emissions in sustainability planning.

Z

Zero-Waste Circular Carbon Economy

A model that eliminates waste, maximizes resource efficiency, and minimizes emissions through closed-loop industrial processes.

Zero Garbage Project – Pune, India

The Zero Garbage Project in Pune, launched in 2012 by the Pune Municipal Corporation (PMC) in collaboration with SWaCH (Solid Waste Collection and Handling) cooperative and Janwani (a social initiative of MCCIA), is a landmark urban sustainability initiative that aims to transform neighborhoods into zero-waste zones through decentralized waste segregation, composting, and community participation.

Appendix I

Overview of Academic Proposals: Theoretical Foundations of Personal Carbon Trading (PCT)

1. **Fawcett, T. (2010)**
    Personal Carbon Trading: A Policy Ahead of Its Time? [53]
    Energy Policy, 38(11), 6868–6876.
    Elsevier
    A foundational work analyzing the feasibility and design challenges of personal carbon trading. This peer-reviewed article offers a comprehensive analysis of the design, feasibility, and challenges of personal carbon trading schemes. It is one of the most cited and influential papers in the field and provides foundational insights that align closely with the eCarbon Card system's goals.

2. **Parag, Y., & Eyre, N. (2010)**
    Barriers to Personal Carbon Trading in the Policy Arena [54]
    Climate Policy, 10(4), 329–345.
    Taylor & Francis

Focuses on institutional, political, and cultural barriers to PCT implementation.

3. **Parag, Y. & Fawcett, T. (2014)**
   Personal Carbon Trading Review: A Review of Research Evidence and Real-World Experience of a Radical idea[55]
   Energy and Emission Control Technologies, 2014(2):23-32
   Dove Medical Press, a part of the Taylor & Francis Group

   Concludes that public acceptability and the cost of the scheme were serious barriers to its introduction. However, a variety of other research work has subsequently demonstrated that public acceptability may not be such a barrier as feared. Nevertheless, there are a number of other barriers, including costs and technical challenges, some adverse distributional effects, and the low carbon capabilities of citizens. Probably <u>the main barrier is the lack of political will currently to consider PCT as a real option</u>. However, opportunities for PCT adoption could open up, particularly if governments fail to meet their carbon reduction targets.

4. **Vandenbergh, M. P., & Gilligan, J. M. (2017)**
   Carbon Governance: Private Governance Responses to Climate Change[56]
   Environmental Law Reporter, 47(2), 10220–10233.
   Environmental Law Institute

   Explores the role of individual and private-sector governance in shaping carbon-conscious behaviors.

5. **Brandon, G., & Lewis, A. (2002)**
   Reducing Household Energy Consumption: A Qualitative and Quantitative Field Study[57]
   Journal of Environmental Psychology, 22(3), 265–276.
   Elsevier

   A behavioral study that informs how feedback and incentives can drive sustained carbon reduction.

6. **Thaler, R. H., & Sunstein, C. R. (2008)**
   Nudge: Improving Decisions About Health, Wealth, and Happiness[58]
   Yale University Press

   Not carbon-specific but essential for understanding behavioral economics and designing choice architecture in systems like eCarbon.

7. **Whitmarsh, L., Seyfang, G., & O'Neill, S. (2011)**
   Public Engagement with Carbon and Climate Change: To What Extent is the Public "Carbon Capable"? [59]
   Global Environmental Change, 21(1), 56–65.
   Elsevier

   Explores the psychological and social readiness of individuals to engage in carbon accountability.

8. **Sorrell, S. (2010)**
Mapping Barriers to Energy Efficiency in the UK
Energy Policy, 38(10), 6123–6139.
Elsevier

A widely cited and highly respected analysis of the economic, institutional, behavioral, and technical barriers that prevent greater uptake of energy efficiency—offering foundational insights relevant to designing equitable and effective personal carbon allowance systems like the eCarbon Card. It explores economic and behavioral constraints on individual-level energy decisions relevant to carbon budgeting.

9. **Capstick, S., & Lewis, A. (2010)**
Personal Carbon Allowances: A Social Psychological Perspective[60]
Climate Policy, 10(4), 371–384.
Taylor & Francis

Investigates public perception and psychological factors affecting acceptance of PCT policies.

10. Anderson, K., & Bows, A. (2011)

Beyond 'Dangerous' Climate Change: Emission Scenarios for a New World[61]

Philosophical Transactions of the Royal Society A, 369(1934), 20–44.

Royal Society Publishing

A seminal and frequently cited work that challenges mainstream emission reduction pathways and calls for more aggressive, equity-based climate strategies. It provides context on why demand-side solutions like PCT are necessary to meet global carbon targets. It aligns strongly with the philosophical underpinnings of the eCarbon Card system — emphasizing urgency, fairness, and transformational approaches.

Appendix II

What Is Blockchain and Why Does It Matter for the eCarbon Card?

What Is Blockchain?

At its core, blockchain is a digital ledger technology — a secure way to record, store, and share data across a network of computers without the need for a central authority.

Think of it as a shared spreadsheet, duplicated across thousands of devices. Every time a new transaction (like a carbon credit trade or deduction) occurs, it gets recorded in a block. These blocks are linked in chronological order, forming a chain — hence the name *blockchain*.

Once data is entered and verified, it cannot be altered, which makes the system tamper-proof, transparent, and trustworthy.

Table 1
Key Features of Blockchain Technology

Feature	Description
Decentralized	No single person or authority controls it. All participants share responsibility.
Immutable	Once data is recorded, it cannot be changed — reducing fraud.
Transparent	Anyone on the network can view the records (if public), promoting accountability.
Secure	Uses advanced cryptography to protect data.

Why Use Blockchain for the eCarbon Card System?

Integrating blockchain into the eCarbon Card system strengthens its credibility, security, and traceability, especially when dealing with personal carbon credits, incentives, and emissions data.

1. **Secure Carbon Tracking**

   Every carbon transaction — whether a purchase, deduction, or trade — is recorded in a transparent and verifiable way, reducing the risk of manipulation or fraud.

2. **Trust Without Central Control**

   Instead of relying solely on a government or corporation, blockchain lets multiple stakeholders (governments, auditors, startups, citizens) verify carbon data independently.

3. **Fair Credit Trading**
   If users trade carbon credits on a marketplace, block-chain ensures that trades are authentic and properly accounted for — building trust between buyers and sellers.

4. **Global Compatibility**
   Blockchain-based carbon systems can eventually link with other countries' platforms, enabling international credit trading, funding, or verification.

5. **Open APIs for Innovation**
   Developers and startups can build on top of the block-chain (e.g., apps for rural users, carbon gamification tools), expanding the system's functionality.

Real-World Examples of Blockchain in Sustainability

- Verra and Gold Standard use blockchain to verify voluntary carbon credits
- Climate Chain Coalition is a global alliance promoting blockchain for climate transparency
- IBM Blockchain is used to trace emissions and carbon offsets in supply chains
- India's own state-level pilots (e.g., for traceable agricarbon credits) are exploring blockchain use for MRV (monitoring, reporting, and verification)

Conclusion

Blockchain isn't about cryptocurrency — it's about trust, transparency, and decentralization. For a nationwide system like the eCarbon Card, which deals with billions of micro-transactions and user identities, blockchain offers a way to keep the system fair, secure, and scalable.

Appendix III

Digital Infrastructure and Technological Framework for eCarbon Card System

This appendix outlines the technological backbone required to support the eCarbon Card system, ensuring it remains accessible, secure, and scalable across India.

1. Digital Platform Design

The eCarbon Card system operates through a centralized platform integrating the following components:

- **Smartphone App and Web Portal:** Allows users to monitor their carbon credits, track emissions, and access reward options in real-time. Features include:

  » Dashboard showing allowance balance.
  » Breakdown of activities contributing to emissions.
  » Recommendations for reducing emissions.

- **Real-Time Data Integration:** Tracks emissions from purchases, energy consumption, and transportation automatically. For example, each fuel purchase or electricity bill is linked to the individual's or business's carbon account.

2. IoT-Enabled Emission Tracking

IoT devices play a vital role in tracking real-time emissions, particularly for businesses. Examples include:

- **Smart Meters:** Monitor energy usage in factories, offices, and homes to calculate carbon emissions.

- **Telematics for Vehicles:** Track fuel consumption and mileage to determine emissions for logistics and personal transport.

- **Waste Monitoring Sensors:** Measure waste generated by businesses to estimate carbon output.

3. Blockchain for Transparency

The system uses blockchain technology to:

- Maintain a tamper-proof ledger of carbon transactions.

- Enhance trust by allowing users to verify their emission records.

- Facilitate seamless trading of carbon credits in the marketplace.

4. Integration with UPI and Digital Payments

Unified Payments Interface (UPI) enables automatic deductions and carbon tracking during financial transactions. Features include:

- Real-time deduction of carbon credits for purchases like fuel, flights, or electricity.

- Incentivized discounts on eco-friendly products purchased via UPI-linked accounts.

5. Data Privacy and Security

Robust encryption protocols ensure data security, while user privacy is safeguarded through anonymized tracking and secure access controls. Policies are implemented to comply with India's data privacy laws.

Appendix IV

Methodology for Allocating Individual Carbon Allowances

The allocation of carbon allowances for individuals is rooted in equity, ensuring inclusivity while aligning with India's climate goals.

1. Data-Driven Approach

Allowances are determined based on:

- **Demographics:** Age, household size, income levels, and geographic location.
- **Lifestyle Patterns:** Urban versus rural living, travel frequency, and energy consumption.

2. Segmentation of Population

To ensure fairness, individuals are categorized into the following groups:

- **Urban Residents:** Lower initial allowances due to higher access to green alternatives (e.g., public transport).
- **Rural Populations:** Higher allowances to account

for limited access to clean energy and modern infrastructure.

- **Special Cases:** Vulnerable groups such as large families, elderly citizens, and individuals with medical needs receive adjusted allowances.

Examples of Allowance Allocation
Table 2

Segment	Baseline Allowance (2026*)	Adjusted Allowance (2035)
Urban Residents	1.5 tons/year	1.2 tons/year
Rural Populations	1.8 tons/year	1.5 tons/year
Vulnerable Groups	2.0 tons/year	1.7 tons/year

*the starting year of the phased implementation of the eCarbon Card system

Income-Based Segmentation
- **Low-Income Group**: Typically, lower-income groups have smaller carbon footprints due to reduced access to high-consumption goods and services. However, their reliance on less efficient fuels (like firewood or coal) can increase emissions. Provide a slightly higher baseline allowance for essential activities, combined with incentives to adopt clean energy alternatives (e.g., subsidies for LPG, solar cooking, or efficient stoves).

» **Allowance:** 1.5 tons per person per year

» **Rationale:** This group typically has lower emissions. They receive a slightly higher allowance to account for essential energy needs, coupled with incentives to switch to cleaner fuels.

- **Middle-Income Group**: This group is growing rapidly in India and often experiences a shift toward higher carbon-intensive lifestyles, such as increased vehicle ownership and consumption of energy-intensive products. Allocate a balanced allowance that promotes sustainable choices, coupled with rewards for energy-efficient practices (e.g., discounts on public transport or incentives for EVs).

  » **Allowance:** 1.3 tons per person per year

  » **Rationale:** Middle-income individuals have moderate consumption and are incentivized to reduce vehicle and energy use. The allowance encourages behavior change towards lower-carbon practices.

- **High-Income Group**: This group has a higher carbon footprint due to greater consumption of goods, frequent air travel, and energy use. Provide a lower baseline allowance, encouraging reductions or purchases of additional credits. Implement stronger incentives for sustainable lifestyles, such as renewable energy installation credits or rebates on eco-friendly vehicles.

» **Allowance:** 1.0 ton per person per year

» **Rationale:** With higher consumption, this group has a larger carbon footprint. They receive a lower allowance to drive reductions and encourage trading or purchasing credits if they exceed it.

Geographical Segmentation

- **Urban Areas**: Urban residents typically have higher emissions due to greater access to electricity, vehicles, and consumer goods. Moderate to lower allowances to encourage reduced energy and transport emissions, supplemented by incentives for using public transit, EVs, and renewable energy.

  » **Allowance:** 1.1 tons per person per year

  » **Rationale:** Urban residents have higher access to public transport and energy-efficient options, so the allowance encourages a shift to public transit and reduced car use.

- **Rural Areas**: Rural populations generally have lower carbon footprints but may rely on biomass for cooking and heating, which contributes to emissions and health issues. Provide a slightly higher allowance for basic energy needs. Pair allowances with incentives for clean energy options, like subsidized LPG, biogas, and solar panel installations.

» **Allowance:** 1.4 tons per person per year

» **Rationale:** Rural residents often rely on traditional fuels. The higher allowance accounts for necessary fuel use, with incentives for adopting cleaner energy.

- **Special Geographical Regions:**
  - **Mountain and Remote Areas**: These regions face limited energy access and higher transportation needs. Allocate additional allowances to account for necessary travel and heating needs. Prioritize green energy access to reduce reliance on fossil fuels.

    » **Allowance:** 1.5 tons per person per year

    » **Rationale:** Due to the need for heating and transportation, remote areas receive a slightly higher allowance, with options to trade credits if they stay under the limit.

  - **High Pollution Zones**: For cities or regions with high pollution (e.g., Delhi NCR), offer lower allowances to encourage faster emission reductions and promote local clean energy alternatives.

» **Allowance:** 0.9 tons per person per year

» **Rationale:** Residents and commuters in major metropolitans are encouraged to seek alternatives which will lower the carbon footprint associated with all their day-to-day activities

3. **Lifestyle and Consumption Patterns**
   - **High Consumption Lifestyles**: Individuals with high-energy consumption, multiple vehicles, or frequent air travel tend to have a larger carbon footprint. Set aa lower initial allowance, incentivizing sustainable behavior. Encourage trading if they exceed their allowance, which could motivate behavior change.

     » **Allowance:** 0.8 tons per person per year

     » **Rationale:** Individuals with high-consumption lifestyles are encouraged to significantly reduce their footprint and purchase credits if they exceed the limit.

   - **Sustainable Lifestyles**: Those already adopting low-energy practices, public transport, and plant-based diets have a smaller footprint. Provide baseline allowances along with additional rewards for consistently staying under limits, motivating them to continue sustainable practices.

» **Allowance:** 1.2 tons per person per year

» **Rationale:** Those already adopting low-carbon behaviors receive a standard allowance along with rewards for continued sustainable practices.

4. Socioeconomic Segmentation
- **Socioeconomic Classes**:
  - **Low Socioeconomic Class**: This group's emissions are typically limited to basic needs. Provide allowances that cover essential emissions, plus targeted subsidies for low-carbon alternatives to traditional cooking and heating fuels.

    » **Allowance:** 1.4 tons per person per year

    » **Rationale:** Higher allowance to meet basic needs, with incentives for adopting cleaner options like solar or biogas.

  - **Middle Socioeconomic Class**: Their emissions increase with lifestyle improvements (e.g., vehicle ownership, more electricity use). Offer balanced allowances to meet needs while promoting lower carbon options, such as public transit or shared mobility credits.

    » **Allowance:** 1.2 tons per person per year

» **Rationale:** Balanced allowance to cover energy and transport needs, with moderate incentives for public transit and efficiency upgrades.

- **Upper Socioeconomic Class**: Higher consumption of luxury goods and services results in a greater carbon footprint. Allocate a lower baseline, encouraging the purchase of additional credits if their footprint exceeds the allowance. Promote luxury sustainable options, like carbon-neutral products or electric vehicles.

  » **Allowance:** 0.9 tons per person per year

  » **Rationale:** Lower allowance to encourage sustainable practices, with trading options for excess usage.

5. **Occupation-Based Segmentation**
   - **Labor-Intensive Occupations** (e.g., agriculture, manufacturing): These occupations might require additional energy or fuel use. Allocate slightly higher allowances for work-related energy needs, with support for low-carbon tools and energy-efficient practices.

   » **Allowance:** 1.4 tons per person per year

   » **Rationale:** Slightly higher allowance due to additional energy needs, with incentives for adopting energy-efficient equipment.

- **White-Collar Occupations**: Often located in urban areas with access to alternative transport options and amenities. Encourage reduced allowances with incentives for using public or shared transport, remote work options, and carbon-neutral commuting choices.

  » **Allowance:** 1.1 tons per person per year

  » **Rationale:** Lower allowance encouraging low carbon commuting options, including public transit, shared mobility, and remote work.

6. **Age and Family Structure**
   - **Families with Dependents**: Larger families may need higher energy consumption. Provide family-based allowances that consider household size, paired with incentives for energy-efficient home appliances and reduced household emissions.

     » **Allowance:** 1.3 tons per family member per year

     » **Rationale:** Family-based allowances account for collective needs, with added credits for energy-efficient appliances and home upgrades.

   - **Young Adults**: Younger individuals may have lower baseline emissions but can benefit from lifestyle-based incentives. Offer standard allowances with rewards for eco-friendly choices and participation in green initiatives.

» **Allowance**: 1.0 ton per person per year

» **Rationale:** Encourages lower consumption and rewards sustainable lifestyle choices.

- **Elderly Individuals:** This group may have lower transportation needs but higher heating and cooling requirements. Provide moderate allowances with subsidies for energy-efficient heating and cooling.

  » **Allowance**: 1.3 tons per person per year

  » **Rationale:** Additional credits for energy-efficient heating or cooling, helping reduce energy use while ensuring comfort.

Implementation and Impact Assessment

1. **Weighted Allowance Calculation**:
    Each citizen is assigned an allowance based on their dominant category within each segment. For example, a middle-income urban resident in a white-collar job with a sustainable lifestyle might receive a combined allowance near 1.1 tons annually.

2. **Projected Emission Reductions**:
    By targeting high-consumption groups with lower allowances and incentivizing trading, India can shift consumption habits and reduce total emissions.

» Assuming that 30% of high-consumption users reduce emissions by 40%, and the rest by 10–15%, a cumulative reduction of 0.7–1.0 billion tons is achievable over time.

3. **Allowance Adjustments and Scaling**:

» Annually, adjust allowances to align with updated targets, ensuring India stays on track to meet 2030 and 2070 goals.

» As infrastructure and green technology adoption improve, reduce allowances progressively, encouraging continued behavior change.

This segmented carbon allowance system recognizes the diversity within India's population, assigning allowances that balance sustainability with fairness. By setting limits based on income, geography, and lifestyle, and enabling a flexible credit-trading market, this system encourages broad participation in India's decarbonization journey. Over time, this approach can significantly lower India's national carbon footprint while fostering sustainable practices across all demographics.

Appendix V

Business Carbon Allowances and Sectoral Benchmarks

This appendix details the methodology for determining annual carbon allowances for businesses, emphasizing sector-specific requirements and benchmarks.

1. Sector-Specific Baselines

Allowances are based on the proportional contribution of each sector to national emissions:

- **Energy Sector:** Higher allowances but steeper reduction targets due to reliance on fossil fuels.

- **Manufacturing and Industry:** Baselines reflect operational scale and production intensity.

    » **Heavy industries (e.g., steel, cement):** Higher allowance initially due to significant energy demands, with annual reductions to align with decarbonization targets.

- **Retail, Logistics, and Transportation:** Modest allowances aligned with sector demands and based on average carbon output and with potential annual reduction path.

- **Agriculture:** Seasonal and resource-intensive processes are considered, with incentives for sustainable practices.

- **Service Sector:** Small to modest allowances due to lower emissions but opportunities for further reductions.

2. Operational Scope of Emissions

Allowances are allocated across:

- **Scope 1:** Direct emissions from operations (e.g., factory emissions). Set allowances for these emissions based on historical data and sector averages.

- **Scope 2:** Indirect emissions from purchased energy – electricity, natural gas etc. Encourage businesses to switch to renewable energy sources to reduce these indirect emissions.

- **Scope 3:** Supply chain emissions – from suppliers, logistics and product usage including transportation, distribution and waste. Provide incentives for companies to reduce these supply chain emissions by partnering with low-carbon suppliers.

3. Production Volume and Efficiency:

- Higher production volumes may warrant higher allowances but are tied to efficiency benchmarks. Encourage sectors to meet energy efficiency standards to qualify for extra credits or allowances.

- Establish benchmarks that reward efficiency. For example, allow companies to earn additional credits if they reduce emissions per unit of production or meet industry efficiency standards.

- Create emissions-to-production benchmarks. For example, emissions per unit of output for manufacturing sectors provide businesses with a clear standard to maintain or reduce emissions relative to production.

4. Benchmarks and Targets

Each sector must achieve annual reductions based on predefined benchmarks. For example:

- **Steel Manufacturing:** Reduce emissions by 5% annually by adopting energy-efficient technologies.

- **Retail Sector:** Transition to renewable energy sources for at least 50% of operations by 2030.

5. Incentives for Exceeding Targets

Businesses exceeding their reduction targets receive tradable credits, fostering competition and innovation within industries. Businesses demonstrating energy efficiency or adopting renewable sources can earn additional credits, encouraging green investments.

Appendix VI

Tracking Carbon Emissions

Carbon cards would track emissions by monitoring various activities and transactions that contribute to an individual's or corporation's carbon footprint. Here's a breakdown of how carbon cards could track emissions across different sources:

1. **Integration with Financial Transactions**
   - **How It Works**: Carbon cards can link with bank accounts or credit card systems to monitor purchases and calculate the carbon footprint of different goods and services. For instance, buying a plane ticket, fuel, or high-carbon foods (like red meat) would add to a person's or organization's carbon tally.

   - **Carbon Calculation**: Each purchase would be assigned a carbon value based on emission factors associated with the product or service. This information could be updated in real-time as each transaction is recorded.

- **Example**: If someone purchases gasoline, the card will automatically add the emissions from fuel combustion based on the amount purchased. Similarly, purchasing an airline ticket could add emissions based on the distance and class of travel.

2. **Utility and Energy Usage Tracking**
   - **How It Works**: Carbon cards could integrate with energy providers to monitor electricity and gas consumption. Utility companies would report the energy usage directly to the carbon tracking system, which would then calculate emissions based on the type of energy source (e.g., coal, natural gas, or renewable).

   - **Carbon Calculation**: Emission factors for electricity and gas are well-established. For instance, using a kilowatt-hour of coal-powered electricity has a higher carbon footprint than the same amount from renewable sources.

   - **Example**: A household using 500 kWh of electricity monthly could have their emissions calculated by multiplying this amount by the emission factor specific to their region's energy mix.

3. Vehicle Emissions Tracking

- **How It Works**: Set up integration with petrol/ diesel pump systems to track fuel purchases or connect to EV charging stations for electric vehicles. Vehicles equipped with telematics or GPS could report mileage and fuel consumption data directly to the carbon tracking system. Alternatively, individuals or corporations could manually report vehicle usage.

- **Carbon Calculation**: Emissions would be calculated based on fuel efficiency and miles driven. This could apply to both personal vehicles and corporate fleets. Based on this calculation, every transaction updates the user's carbon credit balance in real-time, deducting credits based on the fuel or electricity consumed.

- **Example**: A car that consumes 1 gallon of gasoline emits roughly 8.89 kg of CO_2. If a person drives 1,000 miles with a fuel efficiency of 25 miles per gallon, the system will calculate the emissions based on total fuel used.

4. Air Travel and Public Transportation

- **How It Works**: Airlines and public transport systems could share data on passenger travel with the carbon card system. This would allow for automatic tracking of emissions from flights, trains, and buses.

- **Carbon Calculation**: Emissions from air travel depend on flight distance and travel class, while public transport emissions are based on average emissions per passenger-kilometer.

- **Example**: If an individual takes a flight from New York to London, the system could calculate the carbon footprint of that journey based on the emissions per passenger for that distance.

5. **Smart Home and IoT Devices**
   - **How It Works**: IoT (Internet of Things) devices within a smart home can monitor energy usage in real-time. For example, a smart thermostat or appliance could report energy consumption directly to the carbon tracking system.

   - **Carbon Calculation**: Each appliance's energy usage can be multiplied by the regional emission factor for electricity, allowing for granular tracking of household emissions.

   - **Example**: If a smart fridge consumes 100 kWh per month, the carbon card could automatically add the related emissions, helping households identify areas where they can reduce energy use.

6. **Food and Grocery Purchases**
   - **How It Works**: By tracking purchases at grocery stores, the system could estimate the carbon footprint of food based on factors like type (meat, dairy, vegetables) and production methods. Food with a high carbon footprint, like red meat, would add more to the total emissions.

   - **Carbon Calculation**: Carbon footprints for food items are based on lifecycle emissions, from production to transport. For example, 1 kg of beef might emit about 27 kg of CO_2-equivalent, while 1 kg of vegetables emits far less.

   - **Example**: Buying 5 kg of beef per month would add a significant amount to a household's carbon tally compared to purchasing plant-based foods.

7. **Waste and Recycling**
   - **How It Works**: The system could track emissions based on waste generation, encouraging waste reduction and recycling. This might involve tracking data from local waste collection services or recycling centers.

   - **Carbon Calculation**: Waste generates emissions, especially organic waste that decomposes in landfills. Recycling reduces emissions by preventing the need for new material production.

- **Example**: A household that recycles effectively would have a lower carbon tally than one that produces large amounts of landfill waste.

Appendix VII

Integration with UPI and Digital Payments

This appendix provides a detailed overview of how the eCarbon Card system integrates with India's Unified Payments Interface (UPI) to ensure seamless functionality.

- Seamless Integration with UPI and Digital Wallets: Enable users to make transactions directly from their eCarbon Card accounts using Unified Payments Interface (UPI) and popular digital wallets like Paytm or PhonePe. This would require partnership with financial and technology sectors – e.g. collaboration with banks, payment gateways, and digital wallet providers (like Paytm, Google pay, and PhonePe) to integrate carbon tracking features. This integration simplifies transactions, making the card more accessible and user-friendly.

- Automatic Carbon Deductions on Purchases: Partnership with major retailers, fuel stations, and utility providers to allow real-time carbon credit deductions at the point of sale, based on product categories or emissions intensity.

1. Carbon Tracking at Point-of-Sale

Merchants register the carbon intensity of their goods or services with a central database which is interfaced with the UPI system.

When a consumer buys a product through UPI, the UPI system interfaces with this database to retrieve the associated carbon value. UPI transactions automatically deduct carbon credits for purchases based on:

- **Product Type:** Emission intensity associated with goods (e.g., gasoline has higher deductions compared to renewable energy).

- **Service Usage:** Activities such as air travel or public transport.

For example:

- **Fuel Purchase:** A user buys 10 liters of petrol, triggering a deduction of 22 kg CO_2 from their allowance.

- **Flight Booking:** Emissions for a 1,000 km domestic flight (~250 kg CO_2) are calculated and deducted at the time of payment.

QR Code and NFC-Based Transactions

UPI-enabled QR codes and NFC (Near Field Communication) technology can further simplify carbon tracking and deduction:

- **QR Code Scanning for Carbon Data**: Merchants display UPI QR codes that not only facilitate payment but also contain carbon data about the purchase. When the user scans the code, both payment and carbon deduction are processed simultaneously.

- **NFC for Contactless Carbon and Payment Processing**: Using NFC, users can make quick payments with automatic carbon deductions, without needing manual inputs. NFC-enabled carbon tracking could make the process seamless for users.

API Integration for Carbon Tracking and Rewards

An API layer can be built to interface with UPI and the carbon card system, allowing financial institutions, merchants, and carbon-tracking databases to communicate effectively.

- **Carbon Calculation API**: This API calculates the carbon footprint of each transaction based on product details, user preferences, and merchant information.

- **Rewards API**: This API integrates with loyalty programs, credit card rewards, and other incentive systems, enabling carbon credits or "green points" to be awarded to users for staying under their carbon limits or choosing lower-carbon products.

2. Real-Time Notifications

Users receive instant updates on their carbon balance after every transaction, promoting awareness and behavior change.

3. Gamified Rewards and Discounts

- Eco-conscious purchases (e.g., plant-based foods, electric vehicles) earn cashback or credits.

- Retailers participating in the program display carbon ratings for products, encouraging low-carbon choices.

4. Role of Third-Party Data Providers

The system relies on data providers to calculate emission factors for various goods and services. For example:

- **Carbon Footprint Data:** Fuel companies provide emission data per liter of fuel sold.

- **Retail Integration:** Supermarkets link product categories with predefined emission values.

5. Security and Scalability

UPI's proven scalability and robust security infrastructure ensure the system remains resilient against misuse. Blockchain-based tracking further enhances transparency.

Appendix VIII

Integration Across the Value Chain

Here's a detailed roadmap with examples to illustrate how this system could work across the value chain:

1. Raw Material Extraction and Procurement Stage

Carbon Card Integration

At the raw material extraction stage (e.g., mining, agriculture, forestry), companies involved in procuring and extracting raw materials could be assigned carbon quotas based on the expected emissions from their activities. Each entity would have a corporate Carbon Card that records emissions associated with each activity (e.g., deforestation, mining operations, or water and energy use).

Implementation Roadmap

1. **Issue Corporate Carbon Cards**: Assign Carbon Cards to all suppliers of raw materials. For instance, a mining company extracting iron ore for steel production would be issued a Carbon Card.

2. **Track Emissions from Extraction**: Use IoT-

enabled sensors on machinery to monitor energy use and emissions from operations.

3. **Incentivize Low-Carbon Practices**: Reward suppliers who use renewable energy sources or adopt sustainable extraction practices with carbon credits or tax rebates.

Example

Case Study: Renewable vs. Conventional Mining Operations: A company using solar-powered mining equipment would be rewarded with additional carbon credits. These credits could offset their carbon balance, allowing them to stay within their allotted carbon budget and perhaps even sell excess credits to other entities in the supply chain.

2. Manufacturing and Processing Stage

Carbon Card Integration

The manufacturing stage is where raw materials are transformed into intermediate goods or finished products. Factories would have their own corporate Carbon Cards that monitor emissions related to production processes, energy use, and waste.

Implementation Roadmap

1. **Assign Carbon Quotas to Factories**: Each factory receives an annual carbon quota on its Carbon Card, which it must strive to stay within.

2. **Implement Real-Time Carbon Tracking**: Use

IoT devices and smart meters to track emissions from production equipment in real-time.

3. **Encourage Carbon-Efficient Practices**: Offer incentives for factories that use energy-efficient machinery, reduce waste, or adopt circular manufacturing practices.

Example
Case Study: Sustainable Textile Manufacturing: A textile factory producing fabric for clothing could be incentivized to adopt energy-efficient machinery and water recycling systems. By staying within its carbon quota, the factory would receive carbon credits that could be used to reduce its operational costs or sold in a carbon credit marketplace.

3. Distribution and Transportation Stage

Carbon Card Integration
The transportation sector is a significant contributor to emissions, especially in the distribution of goods. Each transport company, whether by land, sea, or air, would have a Carbon Card to monitor emissions related to fuel consumption and travel distances.

Implementation Roadmap
1. **Assign Carbon Cards to Logistics Companies**: Each logistics provider receives a Carbon Card, with emissions tracked based on fuel use, type of vehicle, and distance travelled.

2. **Implement Emission-Tracking Systems**: GPS and telematics can track routes, fuel consumption, and carbon emissions in real-time.

3. **Reward Green Logistics Practices**: Provide incentives for logistics companies using electric or low-emission vehicles, optimizing routes to reduce travel distances, or participating in carbon-offsetting programs.

Example

Case Study: Electric Delivery Fleet: A company using an electric fleet for last-mile delivery would have lower emissions on its Carbon Card. As an incentive, they could earn discounts on road taxes or subsidies for expanding their green fleet, making it more viable for other logistics companies to follow suit.

4. Retail and Consumer Sales Stage

Carbon Card Integration

In the retail stage, stores and e-commerce platforms would track emissions associated with product storage, packaging, and delivery. Additionally, consumers could participate by having individual Carbon Cards that track the carbon impact of their purchases.

Implementation Roadmap

1. **Carbon Cards for Retail Chains**: Large retail chains could be assigned Carbon Cards to monitor emissions from store operations, inventory storage, and supply chain logistics.

2. **Consumer Carbon Tracking**: Introduce consumer-facing Carbon Cards that track emissions associated with individual purchases (e.g., carbon footprint of products bought).

3. **Incentivize Low-Carbon Purchases**: Offer rewards or cashback for consumers purchasing eco-friendly, low-carbon, or recycled products.

Example

Case Study: Carbon-Conscious Consumer Shopping: A supermarket could display carbon ratings for products on shelves. A consumer's Carbon Card app could track the carbon footprint of their shopping basket, offering rewards for choosing products with lower carbon footprints, such as locally sourced or organic items.

5. End-of-Life and Recycling Stage

Carbon Card Integration

At the end-of-life stage, products can either be recycled, repurposed, or disposed of. Recycling companies and waste management facilities would have Carbon Cards that track emissions from their operations and reward effective waste reduction and recycling efforts.

Implementation Roadmap

1. **Assign Carbon Cards to Recycling Plants**: Recy-

cling companies receive Carbon Cards that track emissions from their recycling operations.

2. **Encourage Circular Economy Practices**: Provide incentives for companies that convert waste into new products or raw materials, thus lowering the net carbon impact.

3. **Consumer Involvement in Recycling**: Encourage consumers to participate in recycling by integrating their Carbon Cards with recycling centers, where they can earn carbon credits for responsibly disposing of recyclable items.

Example

Case Study: Electronics Recycling Initiative: Consumers bringing their old electronics to recycling centers could earn carbon credits on their personal Carbon Cards. This promotes recycling of electronics, which often contains toxic substances, and incentivizes sustainable behavior at the consumer level.

Table 3

Stage	Key Emission Source	Carbon Card Mechanism	Incentives & Benefits
Raw Material Extraction	Energy use, deforestation	Emission tracking via sensors	Tax rebates for renewables
Manufacturing & Processing	Industrial energy consumption, waste	Real-time tracking & quotas	Credits for energy-efficient processes
Distribution & Transportation	Fuel consumption, long-distance travel	GPS-based tracking	Discounts on road taxes for EV use
Retail & Consumer Sales	Store energy, packaging waste	Carbon ratings & consumer cards	Cashback for eco-friendly purchases
End-of-Life & Recycling	Waste processing, emissions from disposal	Recycling rewards for consumers	Credits for responsible waste disposal

Summary of Potential Impact

By integrating Carbon Cards across the entire value chain, India can create a comprehensive system for tracking and reducing carbon emissions at every stage of production, from raw material extraction to end-of-life disposal. This system encourages all stakeholders, from suppliers to consumers, to adopt low-carbon practices and make informed choices that collectively contribute to decarbonization goals.

Appendix IX

Alternatives to Blockchain for Carbon Emission Tracking

Here are some alternative methods:

1. **Centralized Database Systems**
   - **How It Works:** A centralized database managed by a trusted authority, such as a government body or international organization, can track carbon emissions and credits. All entities would report their emissions and activities to this central database, where data can be stored, analyzed, and managed by a single administrator.

   - **Pros:** Simple to set up and easy to maintain, with a lower cost than blockchain. Centralized databases allow for better control over data integrity and privacy, and updates can be made more easily.

   - **Cons:** Lack of transparency and potential for manipulation by the central authority. A centralized system may not be as resilient against data tampering and could face cybersecurity risks.

2. Distributed Ledger Technology (DLT) without Blockchain

- **How It Works:** DLT enables the recording of transactions across multiple locations without necessarily using blockchain's consensus mechanisms. Directed Acyclic Graphs (DAGs) are a common alternative to blockchain in DLT.

- **Pros:** DAG-based systems, like IOTA, provide a distributed ledger that doesn't rely on mining, making them more energy-efficient than blockchain. They offer similar transparency benefits and allow for real-time updates across the network.

- **Cons:** Still relatively new and not widely adopted, with fewer established security protocols. DAGs can also become complex to manage, especially in high-scale applications like nationwide carbon tracking.

3. Cloud-based Tracking Platforms

- **How It Works:** Cloud platforms (such as Amazon Web Services, Microsoft Azure, or Google Cloud) provide real-time tracking and analytics solutions that can centralize data from various entities in one location. By leveraging cloud-based tools, it's possible to track, report, and analyze emissions data using standardized protocols.

- **Pros:** Scalable and flexible, with the ability to integrate data from multiple sources easily. Cloud platforms also provide robust security, real-time updates, and advanced analytics tools that can simplify data processing.

- **Cons:** Centralized, which may not ensure transparency or resist tampering by internal actors. Additionally, cloud providers may impose high data storage and maintenance costs, particularly at large scales.

4. **IoT-enabled Tracking Systems with Edge Computing**
   - **How It Works:** Internet of Things (IoT) devices connected to edge computing systems can track emissions data directly at the source, such as in factories or transportation hubs. Data from these devices is processed locally on the edge (near the source), reducing the need for centralized processing.

   - **Pros:** Provides real-time data tracking with low latency, suitable for monitoring emissions directly at the source. Edge computing can reduce bandwidth and storage needs by processing data locally.

- **Cons:** Implementing IoT-enabled tracking requires substantial investment in sensors and infrastructure. Edge computing also has security concerns, as data is stored closer to the source, potentially making it more vulnerable.

5. **Government-regulated Carbon Registries**
   - **How It Works:** Governments can establish carbon registries where companies are required to report their emissions and carbon credit transactions. These registries are maintained by governmental or regulatory agencies, ensuring compliance with carbon reduction targets.

   - **Pros:** Ensures compliance with national and international regulations, offering transparency and accountability. Carbon registries can set clear rules for reporting, verification, and auditing.

   - **Cons:** Limited transparency compared to blockchain, as data can be opaque to the public. Carbon registries are also dependent on regulatory frameworks that may vary by country, creating inconsistencies.

6. **Enterprise Resource Planning (ERP) and Carbon Management Software**
   - **How It Works:** ERP systems like SAP or Oracle can incorporate carbon management modules, allowing

businesses to integrate emissions tracking directly into their operations. Specialized software for carbon management, such as Salesforce Sustainability Cloud or Plan A, can also be used.

- **Pros:** ERP and carbon management software provide comprehensive tools for tracking and analyzing emissions as part of broader business operations. These systems also allow integration with existing enterprise data, making reporting and compliance easier.

- **Cons:** Complex and expensive to implement across multiple industries. Data may be siloed within individual organizations, lacking the transparency of decentralized or open-source solutions.

7. **Third-party Carbon Verification Services**
   - **How It Works:** Third-party organizations, such as Verra or Gold Standard, can verify and certify carbon emissions and offset activities, providing trusted reports for companies and individuals. These certifications can be independently audited and updated as needed.

   - **Pros:** Offers credibility and trust without the need for complex technology. Third-party verification bodies have experience with global carbon standards and can ensure data integrity.

- **Cons:** Manual verification processes may be time-consuming and costly. Lack of real-time updates and difficulty in scaling for small-scale emissions tracking.

8. **Token-based Systems without Blockchain**
    - **How It Works:** A token-based system can be implemented on a centralized platform where carbon credits are issued as "tokens." Unlike blockchain-based tokens, these are tracked in a central ledger maintained by a regulatory body. Each company or individual has an account, and tokens are exchanged digitally.

    - **Pros:** Allows for digital tracking without the complexity of a blockchain. Tokens can be easily tracked, issued, and regulated within a central platform, making the system simple to manage.

    - **Cons:** Requires trust in the central authority to issue, track, and maintain the tokens. Lacks transparency and may be vulnerable to tampering compared to decentralized systems.

9. **QR Codes and Digital Certifications**
    - **How It Works:** QR codes or digital certificates can be issued for specific carbon reduction actions or offsets. Scanning the QR code provides a certificate with details about the carbon credits or reduction measures,

and the certificates can be audited and verified independently.

- **Pros:** Simple to implement and easy to understand for end-users. QR codes can be integrated with online tracking platforms to provide real-time information on carbon credits or offsets.

- **Cons:** Suitable only for smaller-scale tracking, as managing multiple QR codes could become complex. Verification is required to ensure the authenticity of each certificate, which can add administrative overhead.

Table 4

Comparison of Alternatives

Method	Transparency	Cost	Scalability	Complexity	Security
Centralized Database	Low	Low	High	Low	Moderate
DLT without Blockchain	High	Moderate	Moderate	Moderate	High
Cloud-based Platform	Moderate	Moderate-High	High	Low	High
IoT with Edge Computing	Moderate	High	Moderate	High	Moderate

Government Carbon Registry	Low	Moderate	High	Low	Moderate
ERP/Carbon Management Software	Low	High	Moderate	High	Moderate-High
Third-party Verification	Low	High	Low	Low	High
Token-based System (non-blockchain)	Moderate	Low	High	Low	Moderate
QR Codes/ Digital Certs	Low	Low	Low-Moderate	Low	Moderate

Each alternative has its strengths and weaknesses, and the best choice depends on the specific requirements of the carbon tracking initiative. Blockchain provides transparency and security but can be costly and complex, making it less feasible for smaller applications. Centralized databases and cloud-based platforms are affordable and scalable but lack transparency, while IoT with edge computing and ERP systems offer powerful data integration but require significant investment.

For national or corporate-scale tracking, combining several methods—like using a centralized database with IoT sensors and third-party verification—could yield a practical, efficient, and transparent carbon tracking system without relying solely on blockchain.

Appendix X

*Step-by-Step Roadmap for Implementing a Carbon
Card System Across the Value Chain*

1. **Raw Material Extraction and Procurement
   Carbon Card Implementation**:

   - Each supplier of raw materials (e.g., metals, plastics,
     textiles) receives a carbon card that logs on to
     emissions associated with extraction, processing, and
     transportation of raw materials.

   - The carbon card captures data on fuel use, machinery
     emissions, and transportation emissions.

   - Suppliers can reduce their carbon balance by adopting
     greener extraction methods, using renewable energy
     sources, or implementing energy-efficient machinery.

   Example: A mining company extracting lithium for battery production is allocated an annual carbon allowance. They

are rewarded for implementing solar power at mining sites and penalized if emissions exceed their carbon budget.

Case Study: Sustainable Cotton Production: In sustainable cotton production, farmers are encouraged to use organic farming practices and drip irrigation to reduce water and carbon footprints. The carbon card records lower emissions for organic practices, rewarding farmers with incentives.

2. Manufacturing and Processing
Carbon Card Implementation:

- Manufacturers receive carbon cards that track emissions from production processes. Emissions are logged based on energy consumption, waste generation, and transportation.

- Emission benchmarks are set based on industry standards, and companies are incentivized to stay below these limits.

- Carbon credits can be earned by adopting renewable energy, implementing circular economic practices (like recycling), and minimizing waste through lean manufacturing practices.

Example: A car manufacturer uses carbon cards to track the carbon footprint of each stage in the car's production process, from casting metals to assembling components. By using lightweight materials and recycled steel, they reduce their carbon score and earn carbon credits.

Case Study: Carbon-neutral Textile Manufacturing: A textile company shifts to bio-based dyes and renewable energy. Carbon cards track these changes and reward the company by lowering their carbon tax or providing credits, which they can use to invest further in sustainable technology.

3. Distribution and Logistics
Carbon Card Implementation:

- Logistics providers receive carbon cards that track emissions based on fuel consumption, vehicle types, and routes.

- A reduction in carbon emissions can be achieved by optimizing routes, shifting to electric or hybrid vehicles, and using fuel-efficient transportation methods.

- Partnerships with local suppliers are incentivized to shorten the supply chain and reduce transportation emissions.

Example: A food distributor uses electric trucks within urban areas to reduce emissions. The carbon card tracks emissions savings, and the company receives credits that can be used to offset other emissions-heavy practices in their operations.

Case Study: Carbon Tracking in Supply Chains: A retail giant like Amazon could implement carbon cards to measure emissions in each delivery. Partnering with green logistics firms and optimizing warehouses, they lower their carbon score, which could translate to a lower carbon tax or consumer-facing "green delivery" options.

4. Retail and Wholesale
Carbon Card Implementation:

- Retailers have carbon cards that track emissions from store operations, product sourcing, and inventory management.

- Energy consumption from store lighting, heating, and refrigeration is logged.

- Retailers are rewarded for sourcing from low-emission suppliers and implementing energy-saving store designs (e.g., LED lighting, energy-efficient refrigeration).

Example: A supermarket chain earns carbon credits by sourcing locally grown produce, reducing the carbon footprint associated with long-haul shipping.

Case Study: Sustainable Retailing: A retailer like Walmart adopts LED lighting, energy-efficient HVAC systems, and collaborates with eco-friendly suppliers. Their carbon cards log in these efforts, reducing their overall carbon footprint, and potentially passing savings to customers through eco-discounts.

5. Consumer Interaction and Purchase Carbon Card Implementation:

- Consumers are assigned digital carbon cards, tracking the carbon footprint of their purchases.

- At the point of purchase, products carry a carbon score, informing consumers about the environmental impact of their choices.

- Rewards are offered to consumers who opt for low-carbon products, such as discounts, loyalty points, or tax incentives.

Example: A consumer purchases a low-carbon product, such as a sustainably sourced T-shirt. The carbon card logs this purchase, and the consumer receives carbon credits or loyalty points redeemable on future eco-friendly purchases.

Case Study: Consumer Rewards Program: A company like Starbucks could use carbon cards to reward customers who bring their reusable cups or choose plant-based milk options. The carbon card app tracks these choices, rewarding consumers with points redeemable for discounts.

6. End-of-Life Management and Recycling Carbon Card Implementation:

- Manufacturers, retailers, and consumers are incentivized to return products for recycling or proper disposal.

- Carbon cards track the carbon offset achieved through recycling, composting, or upcycling programs.

- Companies that offer take-back programs for end-of-life products can earn additional carbon credits.

Example: An electronics company offers a take-back program for old devices. When customers return their devices, both the company and the consumer receive carbon credits, reducing their overall carbon score.

Case Study: Circular Economy in Electronics: Apple or Samsung could implement a carbon card system where customers return old phones for recycling earn points. The company receives carbon credit, encouraging responsible disposal practices and reducing e-waste emissions.

Appendix XI

Job Creation Estimate

Here's an estimated breakdown of potential job creation across different areas of the economy:

1. **Technology and Infrastructure Development**
   - **Smart Meters and IoT Infrastructure**:
     - » Roles: Installation technicians, maintenance personnel, IoT engineers, and smart grid system integrators.

     - » Job Estimate: With an estimated rollout of smart meters to millions of households, 150,000 - 200,000 jobs could be created in the installation, maintenance, and administration of smart meters and IoT systems.

   - **Blockchain and Digital Infrastructure**:
     - » **Roles**: Blockchain developers, software engineers, data analysts, cybersecurity specialists, and cloud infrastructure managers.

» **Job Estimate**: Developing and maintaining a secure blockchain network and digital carbon credit trading platform would require **50,000 - 75,000 jobs** in specialized tech roles, including blockchain development and cybersecurity.

- **Data Centers**:
  » **Roles**: Data center technicians, facility managers, network engineers, cloud storage experts, and energy-efficient infrastructure specialists.

  » **Job Estimate**: To handle the data generated by millions of transactions and carbon tracking, data centers need expansion, potentially creating **30,000 - 50,000 jobs** in data center management and support roles.

2. **Administrative, Regulatory, and Compliance Roles (Carbon Authority of India)**
   - **Carbon Credit Management and Administration**:
     » **Roles**: Carbon credit administrators, carbon account managers, trading platform administrators, customer service representatives, and transaction auditors.

     » **Job Estimate**: Managing the Carbon Card system and credit trading platform across a population of over a billion people would require **100,000 -**

150,000 jobs in administration, customer service, and transaction management.

- **Government Regulatory Bodies**:
  - » **Roles**: Policy analysts, regulatory compliance officers, data privacy officers, environmental auditors, and legal advisors.

  - » **Job Estimate**: A regulatory authority would be essential to oversee carbon credit trading, allowances, and compliance, creating approximately **20,000 - 30,000 jobs** in policy and regulatory functions.

- **Carbon Auditors and Verifiers**:
  - » **Roles**: Carbon auditors, environmental compliance officers, field inspectors, and environmental impact assessors.

  - » **Job Estimate**: Auditing and verifying emissions reductions for individuals and businesses would generate **50,000 - 70,000 jobs** across environmental compliance and auditing functions.

3. **Manufacturing and Supply Chain**
   - **Smart Meter and IoT Device Manufacturing**:
     - » **Roles**: Assembly line workers, quality control specialists, supply chain managers, and warehouse staff.

» **Job Estimate**: Manufacturing the devices needed for nationwide deployment of smart meters and IoT devices could create **80,000 - 100,000 jobs** in electronics manufacturing and quality assurance.

- **Electric Vehicle (EV) and Renewable Energy Manufacturing**:
  » **Roles**: EV assembly workers, battery technicians, solar panel production workers, and supply chain logistics staff.

  » **Job Estimate**: With increased demand for EVs and renewable energy equipment driven by the carbon credit system, the manufacturing sector could see an additional **100,000 - 150,000 jobs** in EV and solar energy production.

4. **Renewable Energy and Sustainable Projects**
   - **Renewable Energy Infrastructure**:
     » **Roles**: Solar panel installers, wind turbine technicians, renewable energy engineers, and maintenance personnel.

     » **Job Estimate**: Expanding renewable energy infrastructure to support the carbon credit economy would create **200,000 - 250,000 jobs** in installation, engineering, and maintenance of solar and wind energy projects.

- **Carbon Offset and Sequestration Projects**:
  - » **Roles**: Foresters, regenerative agriculture specialists, conservation scientists, project managers, and community outreach coordinators.

  - » **Job Estimate**: Forest conservation, afforestation, and soil carbon sequestration initiatives funded by carbon credits could generate **150,000 - 200,000 jobs** in rural and forestry sectors, including conservation and regenerative agriculture.

5. **Financial and Trading Services**
   - **Carbon Credit Trading Platform**:
     - » **Roles**: Financial analysts, trading platform operators, financial advisors, carbon market analysts, and trading compliance officers.

     - » **Job Estimate**: Developing and operating the carbon trading platform, along with financial advisory services, would create **30,000 - 50,000 jobs** in the financial services sector.

   - **Banking and Financial Institutions**:
     - » **Roles**: Banking associates, digital banking advisors, customer support for carbon credit accounts, and financial planners specializing in carbon trading.

     - » **Job Estimate**: Banks and financial institutions will need support to manage carbon credits as financial assets, creating **20,000 - 30,000 jobs** in banking and finance related to carbon credit services.

6. **Education, Training, and Public Awareness**
   - **Training and Reskilling Programs**:
     » **Roles**: Trainers, educators, program coordinators, environmental consultants, and digital education content creators.

     » **Job Estimate**: Reskilling the workforce for green jobs and ensuring widespread knowledge of the Carbon Card system would create **50,000 - 75,000 jobs** in education, training, and outreach.

   - **Public Awareness and Outreach**:
     » **Roles**: Outreach coordinators, campaign managers, communications specialists, and digital marketers.

     » **Job Estimate**: Public awareness campaigns to educate citizens about carbon credits and sustainable practices would generate approximately **15,000 - 20,000 jobs** in outreach and communications.

7. **Agriculture and Rural Economy**
   - **Regenerative Agriculture and Carbon Farming**:
     - » **Roles**: Farmers trained in regenerative practices, agricultural extension workers, carbon farming project coordinators, and rural program managers.

     - » **Job Estimate**: Implementing carbon farming and regenerative agriculture practices could create **100,000 - 150,000 jobs** in the rural economy, improving agricultural practices and sequestering carbon.

   - **Rural Development Projects and Community Management**:
     - » **Roles**: Community managers, project supervisors, rural outreach coordinators, and eco-tourism guides.

     - » **Job Estimate**: Community-based projects for reforestation, waste-to-energy, and eco-tourism supported by carbon credits could create **50,000 - 70,000 jobs** in rural development.

Table 5
Summary of Estimated Job Creation Across Sectors

Sector	Estimated Jobs Created
Technology and Infrastructure	230,000 - 325,000
Administration, Regulation, Compliance	170,000 - 250,000
Manufacturing and Supply Chain	180,000 - 250,000
Renewable Energy and Sustainable Projects	350,000 - 450,000
Financial and Trading Services	50,000 - 80,000
Education, Training, Public Awareness	65,000 - 95,000
Agriculture and Rural Economy	150,000 - 220,000
Total	**1,195,000 - 1,670,000**

Appendix XII

Calculations

Carbon Sequestration Estimate

- **Average CO2 Absorption per Tree per Year:**
  Approximately 24.6 kg of CO_2[166].

- **Total CO2 Sequestered per Tree over 5 Years:**
  24.6 kg/year × 5 years = 123 kg of CO2.

- **Total CO2 Sequestered by 50 Million Trees over 5 Years:**
  123 kg/tree × 50,000,000 trees = 6.15 billion kg, or 6.15 million metric tons of CO2.

Adjusting for Tree Survival Rates

Considering a conservative survival rate of 20% (accounting for factors like disease, environmental conditions, and maintenance challenges):

- **Effective Number of Surviving Trees:**

  50 million × 20% = 10 million trees.

- **Adjusted Total CO2 Sequestered:**

  123 kg/tree × 10,000,000 trees = 1.23 billion kg, or 1.23 million metric tons of CO2.

Conclusion

If 50 million citizens each plant and nurture one tree annually for five years, and assuming a 20% survival rate, the initiative could offset approximately 1.23 million metric tons of CO2 over the five-year period.

This estimate underscores the significant impact of large-scale tree planting initiatives on carbon sequestration, even when accounting for potential losses.

Additional Resources

1. Carbon Tracking Tools & Calculators

These tools help individuals and organizations measure their carbon footprint, track energy use, and explore reduction strategies.

Online Carbon Footprint Calculators

1. **CoolClimate Calculator (UC Berkeley)**
   » Designed for individuals, households, and businesses.
   » Provides customized reduction strategies.

2. **Carbon Footprint Calculator (CarbonFootprint.com)**
   » Calculates footprints for travel, home energy, food consumption, and lifestyle choices.
   » Offers options for carbon offsetting.

3. **United Nations Carbon Footprint Calculator**
   » Developed by the UN Climate Change Secretariat.
   » Helps track emissions from daily activities and suggests sustainable alternatives.

4. **WWF Footprint Calculator**
   - » Developed by WWF to measure ecological footprint.
   - » Provides lifestyle-based sustainability recommendations.

5. **Global Footprint Network Calculator**
   - » Calculates individual environmental impact using lifestyle data.
   - » Compares footprints to country-wide averages.

6. **EPA Carbon Footprint Calculator**
   - » Designed for U.S. residents.
   - » Focuses on household energy use, transportation, and waste management.

Business & Corporate Carbon Management Platforms

1. **GHG Protocol Corporate Standard**
   - » The most widely used accounting standard for corporate carbon foot printing.

2. **SBTi (Science-Based Targets Initiative)**
   - » Provides guidance for companies to set carbon reduction targets aligned with the Paris Agreement[1].

3. **CDP Carbon Disclosure Platform**
   - » A global disclosure system for businesses to report environmental impact.

4. **Carbon Trust Footprinting Tools**
    » Offers business-level carbon footprint assessment and certification.

5. **Climatiq**
    » API-driven carbon intelligence tool for integrating emissions tracking into business operations.

1. **Mobile Applications for Carbon Tracking & Reduction**

These apps help individuals and businesses track emissions, make eco-friendly choices, and set sustainability goals.

1. **Capture**
    » Tracks daily carbon footprint based on transportation, diet, and energy use.
    » Offers CO_2 reduction goals and behavioral nudges.

2. **JouleBug**
    » Gamifies sustainability by rewarding eco-friendly lifestyle actions.

3. **Earth Hero**
    » Helps users measure, track, and reduce their personal carbon footprint.

4. **Oroeco**
    » Connects spending habits to carbon emissions and offers sustainable alternatives.

5. Sustaina

» Tracks emissions and provides reduction insights for businesses and individuals.

6. Doconomy DO App

» Tracks carbon emissions from financial transactions, helping consumers make low-carbon choices.

7. Giki Zero

» A step-by-step sustainability app to reduce carbon footprints.

8. Klima

» Helps individuals calculate, offset, and reduce their emissions.

9. AWorld (ActNow by the UN)

» Provides tips and challenges for sustainable living.

10. CO2 Living

» Measures carbon impact and integrates with smart home devices for real-time tracking.

Final Note

These resources will serve as practical tools for individuals, businesses, and policymakers to track their emissions, gain insights into decarbonization, and explore scalable climate solutions. By leveraging these tools and further readings, readers of the eCarbon Card System can actively contribute to a sustainable, net-zero future.

Acknowledgments

This book would not have been possible without the inspiration, insight, and encouragement of many.

First, I am grateful to the vast ecosystem of climate scientists, behavioral economists, policy thinkers, grassroots leaders, and sustainability innovators around the world whose ideas have shaped the contours of this vision. The interdisciplinary urgency of decarbonization has been a guiding force, as has the quiet resolve of communities already living within planetary boundaries — often without choice, but always with resilience.

I thank my mentors, collaborators, and colleagues from across academia, startups, industry, and government who have challenged my thinking and helped sharpen the eCarbon Card concept into something bold yet feasible. Your belief in systems thinking, equity, and scalable design kept this work honest and ambitious.

To my family — especially my wife, Dr. Anupama Wahal — thank you for your unwavering support, your patience through long hours of writing, and for anchoring this journey in what truly matters. Your belief in this mission gave me the courage to see it through.

To my brother-in-law, Siddhartha Tandon, cousin, T.N. Kakaji, and friends, Prakash Kota, Kaushal Chari, Manoj

Mishra, Umesh Jayaswal and Rakesh Kaushika for their insightful review and feedback.

To the many readers, changemakers, and policymakers who will carry this idea forward: this book is only the beginning. May it serve as a blueprint, a provocation, and a shared invitation to reimagine climate action — not as sacrifice, but as empowerment.

Thank you.

References

1. https://www.worldometers.info/co2-emissions/
co2-emissions-by-country/

2. https://climatepromise.undp.org/what-we-do/where-we-work/india

3. https://gain.nd.edu/our-work/country-index

4. https://upscgspedia.com/climate-risk-index-2025/

5. https://unfccc.int/process-and-meetings/the-paris-agreement

6. https://en.wikipedia.org/wiki/Aadhaar

7. https://uidai.gov.in/en/

8.https://en.wikipedia.org/wiki/Unified_Payments_Interface

9. https://www.npci.org.in/what-we-do/upi/product-overview

10. https://ondc.org/

11. https://en.wikipedia.org/wiki/
Open_Network_for_Digital_Commerce

12. https://ghgprotocol.org/

13. https://www.iso.org/standard/66453.html

14. https://unfccc.int/ghg-inventories-annex-i-parties/2024

15. https://unfccc.int/process-and-meetings/transparency-and-reporting/reporting-and-review/reporting-and-review-under-the-paris-agreement/national-inventory-reports

16. https://unfccc.int/topics/mitigation/resources/registry-and-data/ghg-data-from-unfccc

17. https://www.cdp.net/en

18. https://en.wikipedia.org/wiki/Carbon_Disclosure_Project#:~:-text=the%20European%20Commission.-,Relevance%20of%20CDP,change%2C%20deforestation%20and%20water%20security.

19. https://www.wri.org/initiatives/science-based-targets

20. https://en.wikipedia.org/wiki/Science_Based_Targets_initiative

21. https://www.e-resident.gov.ee/

22. https://en.wikipedia.org/wiki/E-Residency_of_Estonia

23. https://www.smartnation.gov.sg/

24. https://en.wikipedia.org/wiki/Smart_Nation

25. https://lkyspp.nus.edu.sg/docs/default-source/case-studies/singapores-smart-nation-initiative-final_112018.pdf?sfvrsn=354e720a_2

26. https://www.digilocker.gov.in/

27. https://www.myscheme.gov.in/schemes/onorc

28. https://climate.ec.europa.eu/eu-action/eu-emissions-trading-system-eu-ets_en

29. https://ww2.arb.ca.gov/our-work/programs/cap-and-trade-program

30. https://www.mee.gov.cn/ywdt/xwfb/202407/W020240722528850763859.pdf

31. https://en.wikipedia.org/wiki/Gamification

32. Deci, E. L., & Ryan, R. M. (2000). Self-determination theory and the facilitation of intrinsic motivation, social development, and well-being. *American Psychologist, 55*(1), 68–78. https://doi.org/10.1037/0003-066X.55.1.68

33. https://www.zemo.org.uk/work-with-us/collaborative-initiatives/projects/low-carbon-champions-awards.htm#:~:text=The%20Low%20Carbon%20Champions%20Awards,Business%20Awards%20for%20the%20Environment.

34. https://healthqualitybc.ca/improve-care/low-carbon-high-quality-care/become-a-low-carbon-champion/

35. https://www.the-ies.org/events/low-carbon-champions-awards

36. https://www.theguardian.com/environment/2009/feb/03/personal-carbon-allowances

37. UK's Personal carbon allowances: A revised model to alleviate distributional issues Ecological Economics Volume 130, October 2016, Pages 316-327 (Elsevier)

38. https://www.sciencedirect.com/science/article/abs/pii/S0921800916303354#:~:text=Personal%20Carbon%20Allowances%20(PCAs)%20are,2006%E2%80%932008%20but%20subsequently%20shelved.

39. https://climateledger.org/en/Climate-Credit-Card

40. https://unfccc.int/climate-action/momentum-for-change/activity-database/momentum-for-change-climate-credit-card

41. https://www.cornercard.ch/export/sites/cornercardCH2020/downloads/products_prt/appl_climate_ftrust_en.pdf

42. https://doconomy.com/do-black/

43. https://www.mastercard.com/news/europe/sv-se/nyhetsrum/pressmeddelanden/sv-se/2019/april/do-black-the-world-s-first-credit-card-with-a-carbon-limit/

44. https://www.aspiration.com/zero

45. https://sigmaearth.com/bengal-government-to-launch-worlds-first-carbon-credit-card-for-individuals/

46. https://timesofindia.indiatimes.com/city/kolkata/west-bengal-to-introduce-worlds-first-carbon-credit-cards/articleshow/117318806.cms

47.https://en.wikipedia.org/wiki/Personal_carbon_trading#:~:text=-Carbon%20rationing%2C%20as%20a%20means,when%20buying%20fuel%20or%20electricity.

48. https://www.law.georgetown.edu/environmental-law-review/blog/the-viability-of-personal-carbon-trading/#:~:text=Personal%20carbon%20trading%20refers%20to,an%20individual%20or%20household%20activity.

49.https://www.geos.ed.ac.uk/~sallen/rachel/Climate%20Policy%20Special%20Issue/Fawcett%20and%20Parag%20%282010%29.%20An%20introduction%20to%20personal%20carbon%20trading.pdf

50. https://www.sei.org/projects/personal-carbon-trading-sweden/#:~:-text=The%20efforts%20to%20reduce%20greenhouse,stimulating%20change%20in%20consumer%20behaviour.&text=The%20concept%20of%20personal%20carbon,of%20PCT%20are%20being%20discussed.

51.https://www.sciencedirect.com/science/article/pii/S2213624X21000109#:~:text=Abstract,even%20out%20the%20income%20distribution.

52. https://www.sciencedirect.com/science/article/abs/pii/S0301479723012665

53. https://econpapers.repec.org/article/eeeenepol/v_3a38_3ay_3a2010_3ai_3a11_3ap_3a6868-6876.htm

54.https://www.geos.ed.ac.uk/~sallen/rachel/Climate%20Policy%20Special%20Issue/Parag%20and%20Eyre%20(2010).%20Barriers%20to%20personal%20carbon%20trading%20in%20the%20policy%20arena.pdf

55. https://www.researchgate.net/publication/267324287_Personal_carbon_trading_a_review_of_research_evidence_and_real-world_experience_of_a_radical_idea

56. https://papers.ssrn.com/sol3/papers.cfm?abstract_id=3245243

57. https://colab.ws/articles/10.1006%2Fjevp.1998.0105

58. https://www.researchgate.net/publication/257178709_Nudge_Improving_Decisions_About_Health_Wealth_and_Happiness_RH_Thaler_CR_Sunstein_Yale_University_Press_New_Haven_2008_293_pp

59. https://www.sciencedirect.com/science/article/abs/pii/S0959378010000701

60. https://www.geos.ed.ac.uk/~sallen/rachel/Climate%20Policy%20Special%20Issue/Capstick%20and%20Lewis%20(2010).%20Effects%20of%20personal%20carbon%20allowances%20on%20decision-%20making.pdf

61. https://royalsocietypublishing.org/doi/10.1098/rsta.2010.0290

62. https://www.worldometers.info/co2-emissions/india-co2-emissions/

63. https://unfccc.int/sites/default/files/NDC/2022-08/India%20Updated%20First%20Nationally%20Determined%20Contrib.pdf

64. https://www.impriindia.com/insights/ndc-2023-update-by-india/

65. https://www.unescap.org/sites/default/d8files/28.%20CS-Japan-housing-eco-point-system.pdf

66. https://www.sciencedirect.com/science/article/abs/pii/S0040162521005576?utm_source=chatgpt.com

67. https://ieeexplore.ieee.org/stamp/stamp.jsp?arnumber=10817608

68. https://kar.kent.ac.uk/100231/1/BlockChain%20%26%20Sustainanbility%20PDF.pdf

69. https://www.linkedin.com/pulse/blockchains-role-making-carbon-credit-markets-more-sainath-survase-900ef/

70. https://www.prnewswire.com/news-releases/energy-blockchain-labs-and-ibm-create-carbon-credit-management-platform-using-hyper-ledger-fabric-on-the-ibm-cloud-300425910.html

71. https://www.ccn.com/ibm-develops-blockchain-platform-to-fight-carbon-emissions-in-china/

72. https://blog.energybrainpool.com/en/ibm-and-chinese-energy-blockchain-labs-build-blockchain-based-carbon-asset-manage-ment-platform/

73. https://international.austrade.gov.au/en/news-and-analysis/success-stories/powerledger-creates-the-worlds-first-new-energy-trading-platform

74. https://micropowergrids.com.au/P2P_Energy_Trading/index.html

75. https://depinhub.io/news/powerledgers-peer-to-peer-energy-trad-ing-platform-shows-significant-financial-and-grid-impact-benefits-in-australia-case-study-1394

76. https://www.globalbankingandfinance.com/swytch-and-mobile-bridge-partner-to-drive-clean-energy-generation-globally

77. https://www.ccn.com/swytch-reducing-the-global-carbon-footprint-with-blockchain-technology/

78. https://everledger.io/making-the-commercial-case-for-blockchain-diamond-tracking/

79. https://www.i-diamants.com/en/diamond-traceability-blockchain-Everledger.html,01319?srsltid=AfmBOoqBQ5fIkhJh5Kfah5MlpJJHv-4vsDsoies4haDW_Jq9CRxboB6CM

80. https://www.researchgate.net/figure/Overview-of-the-Everledger-Blockchain-application-in-the-diamond-industry_fig5_339721595

81. https://www.debutinfotech.com/case-study-everledger

82.https://www.google.com/search?q=uk%27s+car-
bon+emission+trading+system+with+blockchain&rlz=1C1YTUH_enU-
S1103US1104&oq=uk%27s+proposed+carbon+emission+trad-
ing+system+with+blockchain&gs_lcrp=EgZjaHJvbWUyBggAEEUY-
OTIGCAEQRRg7MgYIAhBFGDsyBggDEEUYPDIGCAQQRRg8ogEIMzc-
oMGowajeoAgCwAgA&sourceid=chrome&ie=UTF-8

83. https://icapcarbonaction.com/system/files/ets_pdfs/
icap-etsmap-factsheet-99.pdf

84. https://unfccc.int/event/climate-chain-coalition-blockchain-technol-
ogy-for-effective-climate-action-0

85. https://climatechaincoalition.org/

86. https://www.coinsquare.com/en-ca/learn/
un-researches-blockchain-for-climate-change-with-new-coalition

87. https://www.mahindra.com/news-room/press-release/en/mahindra-
and-mahindra-strengthens-its-commitment-to-achieve-carbon-neutrality

88. https://www.mahindra.com/blogs/carbon-neutral-by-2040-science-
based-targets-in-place#:~:text=The%20group%20is%20committed%20
to,science%2Dbased%20targets%2C%20which%20are

89. https://www.tatasteel.com/investors/integrated-report-2022-23/
our-esg-goals.html

90. https://www.unilever.com/news/news-search/2022/
five-ways-were-working-towards-100-renewable-energy-by-2030/

91. https://www.apple.com/newsroom/2020/07/apple-commits-to-be-
100-percent-carbon-neutral-for-its-supply-chain-and-products-by-2030/

92. https://www.dnv.com/maritime/insights/topics/
eu-emissions-trading-system/eu-allowances/

93. https://tyndall.ac.uk/wp-content/uploads/2021/11/twp136.pdf

94. https://www.sciencedirect.com/science/article/abs/pii/
S0921800916303354

95. https://www.sciencedirect.com/science/article/abs/pii/S0301421510005239

94. https://www.flemingpolicycentre.org.uk/personal-carbon-allowances-revisited.pdf

96. https://www.greenstories.org.uk/anthology-for-cop27/solutions/personal-carbon-allowances/

97. http://www.gci.org.uk/Documents/Carbon_Budgets_EAC_Vol_2.pdf

98. Burgess, J., Nye, M., & Whitmarsh, L. (2009). "Governing Individuals' Carbon Emissions: Encouraging Citizen Engagement through Carbon Diaries." *Environment and Planning D: Society and Space*, 27(1), 1–2 https://journals.sagepub.com/doi/10.1068/d4109

99. https://www.gov.uk/government/publications/participating-in-the-uk-ets/participating-in-the-uk-ets

100. https://www.gov.uk/government/news/emissions-scheme-to-reduce-sale-of-carbon-allowances-on-path-to-net-zero

101. https://www.government.se/government-policy/taxes-and-tariffs/swedens-carbon-tax/

102. https://taxfoundation.org/research/all/eu/sweden-carbon-tax-revenue-greenhouse-gas-emissions/

103. https://www.government.se/contentassets/419eb2cafa93423c-891c09cb9914801b/230323-carbon-tax-sweden---general-info.pdf

104. https://academic.oup.com/book/44441/chapter/376663322

105. https://www.sciencedirect.com/science/article/abs/pii/S0961953498000361

106. https://unfccc.int/climate-action/momentum-for-change/planetary-health/alipay-ant-forest

107. https://www.unep.org/news-and-stories/press-release/ant-financial-app-reduces-carbon-footprint-200-million-chinese

108. https://datapopalliance.org/wp-content/uploads/2020/09/Ant-Forest-Report-final-version.pdf

109. https://icapcarbonaction.com/en/ets/china-national-ets

110. https://www.sitra.fi/en/cases/carbon-footprint-calculator/

111. https://e-estonia.com/estonia-to-become-the-greenest-digital-government-in-the-world/#:~:text=Estonia%2C%20known%20for%20its%20digital,digital%20trash%2C%20and%20software%20solutions.

112. https://www.sps.nyu.edu/homepage/metaverse/metaverse-blog/the-estonian-miracle-e-estonia-and-the-future-of-digital-infrastructure.html#:~:text=Thirty%20years%20later%2C%20Estonia%20has%20become%20a,data%20storage%20into%20its%20government's%20day%2Dto%2Dday%20operations.&text=Cybersecurity:%20Estonia%20has%20designed%20its%20blockchain%20ecosystem,data%20are%20secure%2C%20decentralized%2C%20and%20100%%20private.

113. https://e-estonia.com/carbon-footprint-of-estonian-digital-public-services/#:~:text=In%20Estonia%2C%20the%20government%20uses,also%20the%20most%20environmentally%20conscious

114. https://e-estonia.com/estonia-to-become-the-greenest-digital-government-in-the-world/#:~:text=Estonia%2C%20known%20for%20its%20digital,digital%20trash%2C%20and%20software%20solutions.

115. https://www.sei.org/features/partnership-helps-estonian-public-sector-assess-reduce-carbon-footprint/#:~:text=Measurable%20progress%20Results%20from%20last%20year's%20assessment,-consumption%20per%20person%20also%20decreased%20by%2018%.

116. https://icapcarbonaction.com/en/ets/korea-emissions-trading-system-k-ets

117. https://en.wikipedia.org/wiki/Emissions_Trading_Scheme_in_South_Korea

118. https://korea.influencemap.org/policy/-3600bccb14d8b5d92cd3cfb557a8be1e-1830

119. https://time.com/6825324/
south-korea-cap-and-trade-carbon-emissions-polluters-profit-plan15/

120. https://www.spglobal.com/commodity-insights/en/news-research/
latest-news/energy-transition/020725-south-korea-expands-ets-
to-enhance-market-effectiveness-may-cancel-feb-auction-to-curb-
oversupply

121. https://www.nccs.gov.sg/singapores-climate-action/
mitigation-efforts/carbontax/

122. https://www.undp.org/sites/g/files/zskgke326/files/2025-03/
undp-carbon-tax-in-an-evolving-carbon-economy-digital-version.pdf

122. https://www.straitstimes.com/singapore/politics/higher-carbon-
tax-can-spur-green-transition-but-must-be-calibrated-to-help-spore-
firms-stay-competitive

124.https://sustainabledevelopment.un.org/content/documents/1545Cli-
mate_Action_Plan_Publication_Part_1.pdf

125.https://www.dbs.com/newsroom/DBS_LiveBetter_rolls_out_
carbon_calculator_so_customers_can_easily_track_and_offset_their_
carbon_footprint

126. https://www.dbs.com.sg/personal/livebetter

127. https://www.greenfleet.com.au/

128. https://www.joro.app

129. https://theindexproject.org/award/nominees/7797

130. https://www.carbonclick.com/

131. https://www.cogo.co/

132. https://www.ssab.com/en-us/fossil-free-steel

133. https://www.tatasteel.com/investors/integrated-report-2023-24/
climate-change-report.html

134. https://www.tatasteel.com/investors/integrated-report-2018-19/
natural-capital.html#:~:text=Over%20the%20years%2C%20the%20
adoption%20of%20best,efficiency%20as%20well%20as%20reducing%20
carbon%20footprint.&text=This%20could%20reduce%20carbon%20
emission%20by%2050%2D60%25%20as%20compared%20to%20
traditional%20steel%20production.

135. https://consciousplanet.org/en/cauvery-calling

136. https://isha.sadhguru.org/en/wisdom/article/
cauvery-calling-complete-solution-indias-water-crisis

137. https://consciousplanet.org/en/cauvery-calling/campaigns/
cauvery-calling-action-now

138. https://ieefa.org/wp-content/uploads/2021/06/Solar-Powered-Irri-
gation-Would-Accelerate-Indias-Energy-Transition_June-2021.pdf

139. https://www.afdb.org/en/topics-and-sectors/
initiatives-partnerships/sustainable-energy-fund-for-africa

140. https://mahindralogistics.com/transitioning-to-electric-vehicles/

141. https://mahindralogistics.com/logiedel-ignites-success-with-90-
reduction-in-vehicle-breakdowns-and-30000-kg-co2-saved-monthly/

142. https://india.mongabay.com/2021/04/
kochis-disappearing-mangroves-lock-away-significant-carbon/

143. https://gulfbusiness.com/uaes-etihad-launches-conscious-choices-
worlds-first-green-loyalty-programme/

144. https://www.linkedin.com/pulse/
eco-friendly-uae-corporate-sustainability-success-stories-js-kho-yaksf/

145. https://www.csrbox.org/
India_CSR_Project_KPIT-Technologies-Ltd--Maharashtra_6071

146. https://www.persistentfoundation.org/archives/
zero-garbage-project/

147. https://www.weforum.org/projects/project-zero-waste/#:~:text=According%20to%20the%20data%20by,reusing%2C%20recycling%20or%20proper%20disposal.

148. https://satsang-foundation.org/mytree-project/mytree-projects-contribute-to-greening-efforts-across-the-country/

149. https://www.weforum.org/stories/2024/04/bengaluru-saytrees-environmental-restoration-movement/

150. https://www.ikea.com/us/en/newsroom/corporate-news/ingka-group-accelerates-investments-in-renewable-energy-to-exceed-2020-target-puba464c247/

151. https://www.ikea.com/global/en/our-business/sustainability/our-circular-agenda/#:~:text=Designing%20all%20products%20with%20circular%20capabilities&text=This%20means%2C%20we%20design%20for,reassembly%2C%20remanufacturing%2C%20and%20recyclability.

152. https://www.ikea.com/sg/en/this-is-ikea/climate-environment/climate-action-pub85dbcef0/#:~:text=Becoming%20climate%20positive,-This%20means%20reducing&text=We%20are%20committed%20to%20reducing,by%202050%20at%20the%20latest.

153. https://www.infosys.com/about/corporate-responsibility/documents/pioneering-net-zero-buildings.pdf

154. https://www.infosys.com/sustainability/documents/esg-2022-23/env-story1.pdf

155. https://www.reccessary.com/en/news/sg-regulation/singapore-firms-turn-to-ev-clean-cookstoves-carbon-credits-from-ghana#:~:text=Singapore%20firms%20turn%20to%20EV%2C%20clean%20cookstoves%20carbon%20credits%20from%20Ghana,-Regulation%20August%2026&text=To%20reduce%20carbon%20tax%20expenditures,time%20to%20prove%20their%20effectiveness.

156. https://wwp.org.br/wp-content/uploads/02.-Bolsa-Verde-Program-Sheet.pdf

157. https://www.ilo.org/resource/news/
jobs-renewable-energy-record-highest-annual-growth-rate-reaching-162

158. https://pib.gov.in/PressNoteDetails.
aspx?NoteId=153238&ModuleId=3&reg=3&lang=1

159. https://www.tice.news/tice-dispatch/business-finance-economy-
and-corporate-news/how-are-green-jobs-shaping-indias-future-work-
force-and-economy-7605979

160. https://indiaghgp.org/

161. https://ieefa.org/sites/default/files/2022-10/Indian%20Resi-
dential%20Rooftops-%20A%20vast%20Trove%20of%20Solar%20
Energy%20Potential_Oct2022.pdf

162. https://www.weforum.org/stories/2019/02/the-netherlands-is-
giving-tax-breaks-to-cycling-commuters-and-they-re-not-the-only-ones/

163. https://www.unilever.com/our-company/
our-history-and-archives/2010-2020/

164. https://greennetwork.asia/featured/how-unilever-transforms-its-
business-with-unilever-sustainable-living-plan/#:~:text=Unilever%20
Sustainable%20Living%20Plan%20(USLP)%20was%20launched%20
in%202010.,save%20costs%2C%20and%20fuel%20innovation.

165. https://en.wikipedia.org/wiki/Global_Carbon_Project

166. https://www.fortomorrow.eu/en/blog/co2-tree?

Bibliography and Further Reading

Comprehensive list of references used in the development of the book, including academic papers, reports, and credible sources on sustainability and climate policy.

Books on Climate Change and Decarbonization

1. *How to Avoid a Climate Disaster* – Bill Gates, Knopf (2021, New York)
    - A practical guide on achieving net-zero emissions through innovation and policy.

2. *Drawdown: The Most Comprehensive Plan Ever Proposed to Reverse Global Warming* – Paul Hawken, Penguin Books (2017, New York)
    - A scientific analysis of scalable climate solutions.

3. *The Uninhabitable Earth: Life After Warming* – David Wallace-Wells, Tim Duggan Books (2019, New York)
    - A deep dive into the catastrophic risks of climate inaction.

4. *Speed & Scale: An Action Plan for Solving Our Climate Crisis Now* – John Doerr, Penguin Portfolio (2021, New York)
   - A data-driven roadmap for achieving net-zero emissions.

5. *The Sixth Extinction: An Unnatural History* – Elizabeth Kolbert, Henry Holt and Company (2014, New York)
   - Examines the impact of human-driven climate change on biodiversity.

6. *The Big Fix: Seven Practical Steps to Save Our Planet* – Hal Harvey & Justin Gillis, Simon & Schuster (2022, New York)
   - Offers pragmatic solutions for global decarbonization.

7. *The Future We Coose: The Stubborn Optimist's Guide to the Climate Crisis* – Christiana Figueres & Tom Rivett-Carnac, Knopf (2020, New York)
   - A hopeful yet urgent guide showing how our choices today shape a livable or catastrophic climate future - and empowers readers with mindset shifts and actions for a livable planet.

8. *Planetary Economics: Energy, Climate Change, and the Three Domains of Sustainable Development* – Michael Grubb, Routledge (2014, Abingdon, UK)
   - A comprehensive look at energy policy and climate mitigation strategies.

Scientific Reports & Whitepapers

1. **IPCC Climate Reports** – "Climate Change 2023: Synthesis Report" (2023) – https://www.ipcc.ch/report/ar6/syr/
   - The leading scientific assessment on climate change from the Intergovernmental Panel on Climate Change (IPCC).
     1. McKinsey Report: The Net-Zero Transition – "The Net-Zero Transition: What It Would Cost, What It Could Bring" (2022) – https://www.mckinsey.com/business-functions/sustainability/our-insights/the-net-zero-transition
        - A comprehensive assessment of economic changes required to achieve net-zero emissions.
     2. World Economic Forum (WEF) Climate Initiatives – "Climate Action Platform: Strategic Intelligence Brief" (2022) – https://www.weforum.org/agenda/archive/climate-change/
        - Policy recommendations and industry insights for sustainability.
     3. IEA Net Zero by 2050 Report – "Net Zero by 2050: A Roadmap for the Global Energy Sector" (2021) – https://www.iea.org/reports/net-zero-by-2050
        - A global roadmap for the energy sector to reach carbon neutrality.

4. Project Drawdown Climate Solutions – "Drawdown: The Most Comprehensive Plan Ever Proposed to Reverse Global Warming" (2017) – https://www.drawdown.org/solutions
   - Research-based ranking of top climate mitigation solutions.

5. UNEP Emissions Gap Report – "Emissions Gap Report 2023" (2023) – https://www.unep.org/resources/emissions-gap-report-2023
   - Annual assessment of global emissions reduction efforts.

6. Climate Transparency Report – "Climate Transparency Report 2023: Comparing G20 Climate Action Towards Net Zero" (2023) – https://www.climate-transparency.org/g20-climate-performance/g20report2023 A comparative analysis of G20 countries' climate actions.

2. Government & Policy Resources

- **Ministry of Environment, Forest and Climate Change (MoEFCC), Government of India. (2008).** *National Action Plan on Climate Change (NAPCC).* **https://moef.gov.in/en/division/environment-divisions/climate-changecc-2/**
   » India's official climate policy framework.

- **European Commission. (2020).** *The European Green Deal.* **https://ec.europa.eu/clima/eu-action/ european-green-deal_en**
  - » The European Union's roadmap for achieving carbon neutrality.

- **U.S. Inflation Reduction Act (Climate Policy) – The White House. (2022).** *Clean Energy and Climate Provisions in the Inflation Reduction Act.* **https://www. whitehouse.gov/cleanenergy/**
  - » U.S. climate and clean energy legislation details.

3. **Climate Science & Global Frameworks**

- **Intergovernmental Panel on Climate Change (IPCC). (n.d.).** *IPCC Reports.* **https://www.ipcc.ch**
  - » United Nations Framework Convention on Climate Change (UNFCCC). (n.d.). *Official Portal.* https:// unfccc.int
  - » Hawken, P. (Ed.). (2017). *Drawdown: The most comprehensive plan ever proposed to reverse global warming.* Penguin Books.
  - » Rockström, J., & Gaffney, O. (2021). *Breaking boundaries: The science of our planet.* DK Publishing.

4. Carbon Markets & Policy Instruments

- **World Bank. (2023).** *State and Trends of Carbon Pricing 2023.* **https://www.worldbank. org/en/topic/climatechange/publication/ state-and-trends-of-carbon-pricing-2023**
  - » Organigation for Economic Co-operation and Development (OECD). (2022). *Carbon pricing and energy taxation: How to get carbon prices right.* https://www.oecd.org/tax/tax-policy/carbon-pricing-and-energy-taxation.htm
  - » Metcalf, G. E. (2019). *Paying for pollution: Why a carbon tax is good for America.* Oxford University Press.
  - » European Commission. (n.d.). *EU Emissions Trading System (EU ETS).* https://climate.ec.europa.eu/eu-action/ eu-emissions-trading-system-eu-ets_en

5. Behavioral Economics & Incentives

- **Thaler, R. H., & Sunstein, C. R. (2009).** *Nudge: Improving decisions about health, wealth, and happiness.* **Penguin Books.**
- **Ariely, D. (2008).** *Predictably irrational: The hidden forces that shape our decisions.* **HarperCollins.**
- **Jenks, B., Vaughan, P. W., & Butler, P. J. (2021).** *Making behavior change stick: Insights from conservation and sustainability.* **Rare. https://rare.org/ report/making-behavior-change-stick/**

- **Digital Public Infrastructure & Innovation**
- **Nilekani, N., & Shah, V. (2015).** *Rebooting India: Realizing a billion aspirations.* **Penguin Random House India.**
- **Unique Identification Authority of India (UIDAI); National Payments Corporation of India (NPCI); Open Network for Digital Commerce (ONDC). (n.d.).** *Official Portals.* **https://uidai.gov.in, https://npci.org.in, https://ondc.org**
- **Digital Public Goods Alliance. (n.d.).** *Reports and publications.* **https://digitalpublicgoods.net**
- **Sustainable Development & Equity**
- **The Energy and Resources Institute (TERI). (n.d.).** *Annual energy and environment reports.* **https://www.teriin.org**
- **Council on Energy, Environment and Water (CEEW). (n.d.).** *Publications.* **https://ceew.in**
- **Sen, A. (1999).** *Development as freedom.* **Knopf.**
- **Raworth, K. (2017).** *Doughnut economics: Seven ways to think like a 21st-century economist.* **Chelsea Green Publishing.**

6. India-Specific Climate & Governance Literature

- **NITI Aayog. (2018).** *Strategy for New India @75.* **https://www.niti.gov.in**
- **Ministry of Environment, Forest and Climate Change (MoEFCC), Government of India. (n.d.).** *India's Biennial Update Reports to UNFCCC.* **https://moef.gov.in**

- **World Resources Institute (WRI) India. (n.d.).** *Reports on urban sustainability and energy transition.* **https://wri-india.org**

7. Global Case Studies & Inspirations

- **Figueres, C., & Rivett-Carnac, T. (2020).** *The future we choose: Surviving the climate crisis.* **Knopf.**
- **World Economic Forum. (2021).** *Net zero carbon cities: An integrated approach.* **https://www.weforum.org/reports/ net-zero-carbon-cities-an-integrated-approach**
- **United Nations Development Programme (UNDP). (2020–2023).** *Human Development Reports.* **https:// hdr.undp.org**
- **Case studies from national programs: South Korea ETS, California Cap-and-Trade, Japan's Eco-Points Program. (Various).** *Referenced in comparative policy literature.*

Author Bio

Sanjay Wahal, PhD, is a seasoned sustainability strategist, systems thinker, and
climate innovator with deep expertise in
decarbonization technologies, bio-based
materials, and circular economy solutions.
With a doctorate in Chemical Engineering, an MBA in Strategy and Innovation,
undergraduate degree from IIT Kanpur,
and a career spanning over three decades of cross-functional
experience, Dr. Sanjay Wahal brings rare clarity to complex
global challenges. His career spans industry leadership, applied
research, and strategic advisory roles across climate innovation,
advanced clean technologies, and the transition to low-carbon
pathways and regenerative systems. He has advised startups,
Fortune 500 companies, and institutional investors worldwide,
and is the author of multiple technical publications and co-in-
ventor on several patents.

Dr. Wahal is the founder and president of Decarbonization, LLC, an advisory and consulting firm at the intersection of
science, AI/ML, policy, and market transformation. His work
empowers institutions to reimagine carbon not just as a liability,
but as a strategic and societal asset. He has collaborated with

leading climate-tech ventures and impact investors to accelerate decarbonization through innovation, finance, and inclusive strategy. Dr. Wahal is the lead author of *"Forecasting a United States Carbon Allowance Price Index Using Neural Networks and Macroeconomic Indicators,"* currently under review for publication in a reputed journal.

With a unique blend of technical depth, strategic insight, and intimate understanding of India's digital and policy landscape, Dr. Wahal brings both grounded realism and visionary clarity to the eCarbon Card concept. His frequent engagements with policymakers, corporate leaders, and civil society further enriches this book's actionable, citizen-first approach to climate action.

He and his wife, Anupama, live in Appleton, WI.

To learn more, visit www.decaerbonizationllc.com
and www.sanjaywahal.com

www.ingramcontent.com/pod-product-compliance
Lightning Source LLC
Chambersburg PA
CBHW021351210326
41599CB00011B/832